EMMA WOOD was born in Yorkshire in 1958. She stayed on in East Anglia after reading History at Cambridge University. She did a variety of jobs after graduating and finished up as a partner on a junk stall which traded on markets throughout the area. She moved from Norfolk to Sutherland in 1987 and after two years went to Ross-shire. She lives on the Heights of Achterneed where she gives English lessons and does freelance writing and editing. Her book *Notes from the North: Incorporating a Brief History of the Scots and the English* was published in 1998.

Emma's interest in the modernisation of the Highlands led her to the history of hydro power developments in the north. She found this 'a totally fascinating subject', and one with links to a range of crucial topics.

She has one daughter Jasmine.

The Hydro Boys

Pioneers of Renewable Energy

EMMA WOOD

Luath Press Limited

EDINBURGH

www.luath.co.uk

First Published 2002
Reprinted 2002
Revised Edition 2004

The paper used in this book is recyclable. It is made from low chlorine pulps produced in a low energy, low emission manner from renewable forests.

Printed and bound by
Bookmarque Ltd., Croydon

Map by Jim Lewis

Typeset in Sabon by S. Fairgrieve, Edinburgh 0131 658 1763

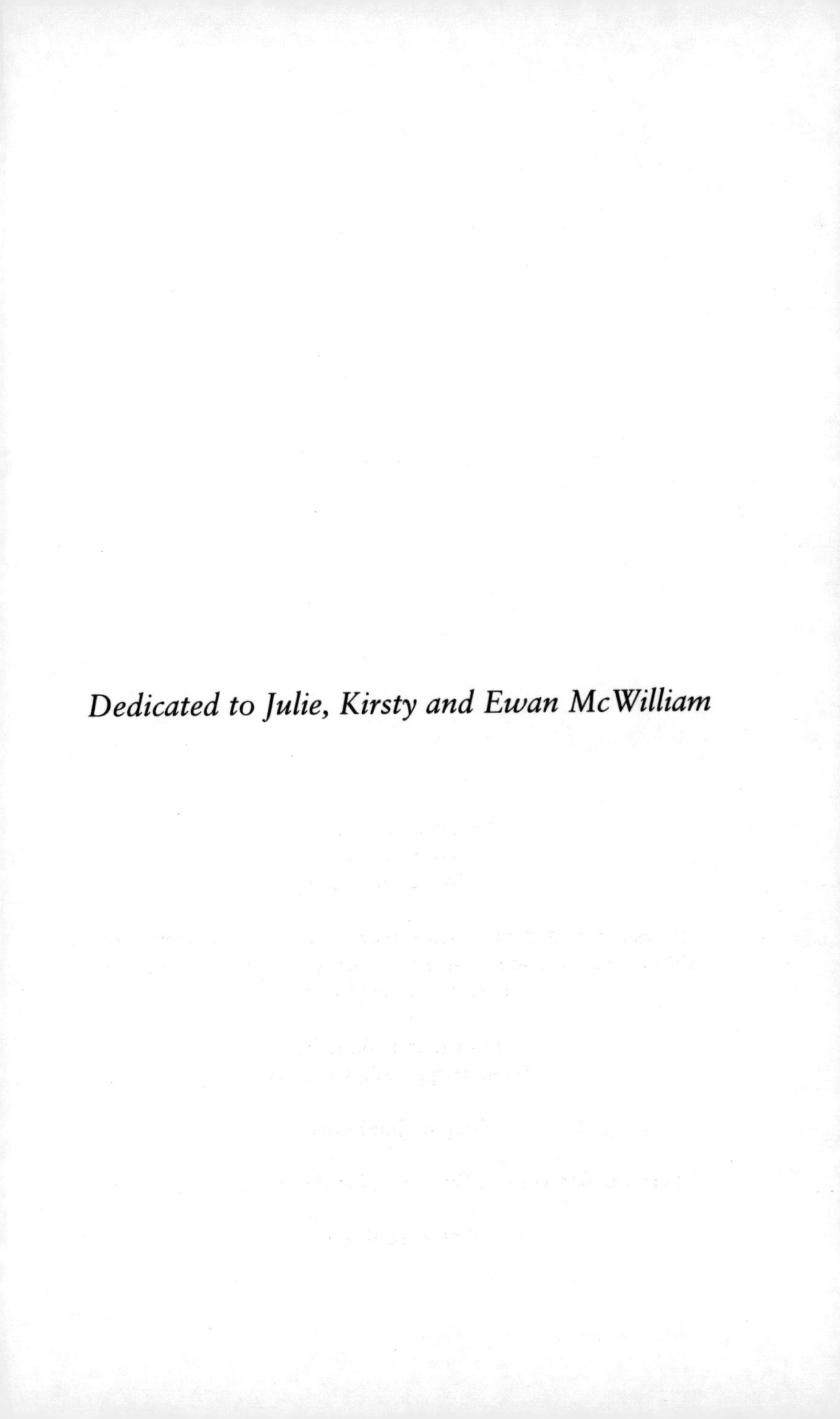

Dedicated to Julie, Kirsty and Ewan McWilliam

Acknowledgements

This book would simply never have existed without the help of the people named below. They gave me their time, expertise and memories and contributed an inspirational wealth of information and insights without which *The Hydro Boys* could never have been written. The length of the list and the depth of my gratitude to its members must mean that all I did was work the word processor: the real authors, apart from the ones who preferred to remain anonymous, are listed here. My heartfelt thanks go to them all including Patrick Agnew, James Black, Andrew Macleman, Wendy de Russet, Tom Douglas, Helen Fullarton, Jessie Harrow, James Hunter, Geordie Macintyre, Ian MacLeod, Hugh and Amelia Maccoriston, Alison MacDonald, Campbell Macdonald of Scottish and Southern Energy, Kyo Mackay, Iain Mackay, John Mackenzie, Ernie Dan Mackenzie, Roy Mackenzie, Ina Mackillop, Willie Maclennan, Angus MacPhee, Willie Ross, AD Ross, Scott Russell, Bob Sim, Douglas and Alison Watson. I take full responsibility, however, for the book's contents.

As well as these invaluable interviewees I must also thank everyone who has helped and encouraged me in the writing of *The Hydro Boys*. My special thanks go to Edinburgh Friends of the Earth, the staff at Dingwall Library, Julia Russell, Dorothy Burr, Lizbeth Collie, Jem Taylor, Lucy Dargue, Marj Donaldson, Julie McWilliam, Rod and Beth Harbinson, Sean Wood, Rod and Pat Richard, Fiona Robertson, Nick Rochford, Douglas Williamson, Jasmine Woodcraft, Dawn Dalgetty, the Dingwall Community Nurses, the staff of Ward 7A, Raigmore Hospital and the late Dougie Chalmers.

I am deeply grateful to the Scottish Arts Council for the Writer's Bursary which enabled me to undertake this project and to Scottish and Southern Energy plc for permission to include NOSHEB photographs.

For permission to reproduce prose extracts the publisher gratefully acknowledges the following:

Scottish and Southern Energy plc for all extracts from *The Hydro* by Peter Payne;

Mainstream Publishing for the extract from *Last of the Free* by James Hunter;

House of Lochar for the extract from *Off in a Boat* by Neil Gunn;

Birlinn Ltd for the extracts from *Children of the Dead End* by Patrick MacGill and *Isolation Shepherd* by Iain Thomson.

All attempts have been made to trace the copyright of other passages and pictures used. The author will be pleased to include any copyright holder who wishes to be acknowledged in any future edition of this work.

For their generous assistance in helping me prepare the paperback edition I would like to thank Tom Douglas, Peter Brookes, Steve Evans, Una Lee, Andrew Macleman, Bob Sim, Neil Sandilands of Scottish and Southern Energy plc and Maf Smith of the Scottish Renewables Forum.

Contents

Preface to paperback edition of *The Hydro Boys*

THE HARDBACK EDITION of *The Hydro Boys* ended on an expectant note as opportunity seemed to be knocking for Scottish hydro. The three years which have passed since I finished writing it have seen exciting developments in hydro and the entire renewables field. However, these developments have occurred against a background of serious deficiencies in Scottish and UK energy policies. This edition includes an examination of these deficiencies and the remedies that could ensure hydro power and the rest of our renewable energy resources are utilised to their full generating and environmental potential.

Emma Wood
March 2004

Preface

THE SECOND MILLENNIUM AD saw a gradual decline in the independent strength of Highland society, a decline that intensified disastrously with the Jacobite debacle and the Clearances. Remnants of the old culture survived in the north well into the last century, albeit marginalised and impoverished. As a newcomer to the Highlands fifteen years ago, I set about discovering as much as I could about the region's history. I was fascinated by the remaining traces of pre-industrial tradition which had been swept away elsewhere in the British Isles. Today, the Highlands and Islands seem very modern: cosmopolitan, fashionable and on-line. I became just as intrigued by this sea-change from ancient to modern as I had first been by earlier Highland history. When exactly did this huge change of Highland content and mood take place and what triggered it off?

Explanations of historical change rarely hinge on single factors but the pioneering work of the North of Scotland Hydro-Electric Board certainly coincided with the unprecedented modernisation of the Highlands that followed the Second World War. In 1960, with much of the NOSHEB construction programme complete, the area was set firmly on a course of progress that would have been unthinkable without the provision of affordable electricity for the bulk of the Highland population. Beginning in the late 1940s with the arrival on Highland farms of electrical grain-drying equipment, the use of technology to help the Highlands catch up economically with the rest of the British Isles has come a long way. The logical conclusions are still unfolding, including the arrival of computerised higher education, tourism and call-centres: all impossible without the pioneering work of the Hydro Board.

It seemed that the pivotal importance of the Board in post-war Highland history deserved further investigation. When I asked questions about hydro power in the Highlands, I received consistently fascinating answers. I heard about drowned farms and hamlets, the ruination of the salmon-fishing and how Inverness might be washed away if the dams failed inland. I was told about the huge veins of crystal they found when they were tunnelling deep under the mountains and when I wanted to know who 'they' were: what stories I got in reply! I heard about Poles, Czechs, poverty-stricken Irish, German spies, intrepid locals and the heavy drinking, fighting and gambling which went on in the NOSHEB contractors' camps.

The politics turned out to be enthralling too, with an excellent cast of goodies and baddies and some unexpected outcomes. Winston Churchill appears in the unlikely role of chief political sponsor of Britain's first nationalised industry and then there's the fierce national debates which raged all through the history of hydro power development in the Highlands. Questions of community land rights, sustainable energy, planning policy and landscape preservation cartwheel through a century which started with the dawn of the electric age and ended with Climate Change threatening to put the lights out all over the developed world.

Most extraordinary of all was the absence of any public recognition in the Highlands of the Board's achievement or of the importance of its political godfather, Thomas Johnston. There are no monuments to him or the men who designed and built the hydro power schemes anywhere in the Highlands. Even the last post-privatisation reminder of the old Hydro Board, the name Scottish Hydro plc, is being replaced by Scottish and Southern Energy's corporate logo on the company's Highland van fleet.

The generations responsible for the NOSHEB achievement are not getting any younger. They are witnesses to a unique project and their recollections are accordingly fascinating and invaluable. Their stories have a very special role to play in the modern history of the Scottish landscape. From 1745 to the coming of the NOSHEB, that landscape was the sole preserve of defeated clans, exploited tenants and the sportsmen and tourists whose presence signalled the appropriation of the Highlands by outsiders. To focus on the hydro boys hard at work in the Highland hills makes a glorious change: they were the first people for many a year to be striving freely in that landscape for their own benefit and for the benefit of the Highland people. Perhaps for me that was the supreme attraction of the hydro story: it makes such a great corrective to the Highland landscape's sense of desolate emptiness, which otherwise grows more oppressive with each page of Highland history.

Hydro power has a promising future as well as a fascinating past. I soon became enthralled by the idea of human ingenuity making electricity from rainfall and gravity: perfectly renewable resources. I love the idea of the Highlands being the site of such clever and useful technology. Like many before me I have become a hydro power fanatic, and discovered a great story at the same time.

I must warn readers not to expect an encyclopaedic account of Highland hydro power or the NOSHEB project. To avoid repetition, the material which follows does not discuss every Highland hydro scheme individually. Rather, its content reflects the knowledge and experiences of those who were kind enough to tell me their part of the story.

Emma Wood
April 2002

Thus it came about that one of my earliest memories is of the evening, not far into the 1950s, when my sister and I were able to greet our father's homecoming from work by throwing the switch, which miraculously as it seemed to us, filled our Duror kitchen with what we called, for years afterwards, electric light.

James Hunter
Last of the Free

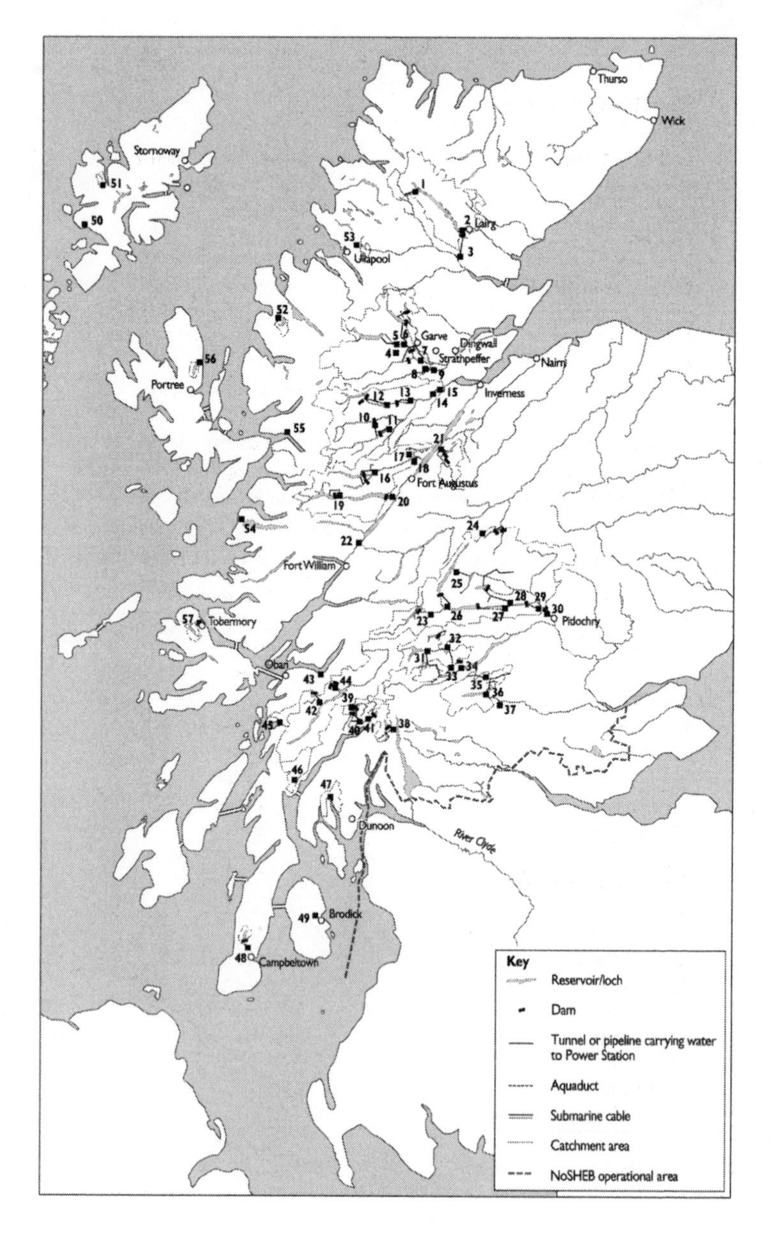

Thurso
Wick
Stornoway
51
50
1
2
Lairg
3
53
Ullapool
52
Garve
Dingwall
Strathpeffer
Nairn
56
Portree
Inverness
55
Fort Augustus
54
Fort William
57
Tobermory
Pitlochry
Oban
Dunoon
River Clyde
Brodick
Campbeltown
Key
Reservoir/loch
Dam
Tunnel or pipeline carrying water to Power Station
Aquaduct
Submarine cable
Catchment area
NoSHEB operational area

POWER STATIONS

Loch Shin Scheme
1 Cassley
2 Lairg
3 Shin

Conon Valley Scheme
4 Achanalt
5 Grudie Bridge
6 Mossford
7 Luichart
8 Orrin
9 Torr Achilty

Affric-Beauly Scheme
10 Mullardoch
11 Fasnakyle
12 Deanie
13 Culligran
14 Aigas
15 Kilmorack

Garry-Moriston Scheme
16 Ceannacroc
17 Livishie
18 Glenmoriston
19 Quoich
20 Invergarry

Tummel-Garry Scheme
23 Gaur
24 Cuaich
25 Loch Ericht
26 Rannoch
27 Tummel
28 Errochty
29 Clunie
30 Pitlochry

Breadalbane Scheme
31 Lubreoch
32 Cashlie
33 Lochay
34 Finlarig
35 Lednock
36 St. Fillans
37 Dalchonzie

Sloy & Awe Schemes
38 Sloy
39 Sron Mor
40 Clachan
41 Allt-na-Lairige
42 Nant
43 Inverawe
44 Cruachan
45 Kilmelfort
46 Loch Gair
47 Striven
48 Lussa

Other Stations
21 Foyers
22 Mucomir
49 Brodick
50 Chliostair
51 Gisla
52 Kerry Falls
53 Loch Dubh
54 Morar
55 Nostie Bridge
56 Storr Lochs
57 Tobermory

Introduction

THE NORTH OF SCOTLAND Hydro-Electric Board's Development Plan was carried out from 1943 to 1965. This was a truly heroic episode in a Highland history dominated for the previous two centuries by the Clearances and their aftermath of neglect and disadvantage. The story of hydro power in the Highlands is packed with brave and determined characters, from the late nineteenth century pioneers of hydro power development and their ragged workforce, to the designers, planners and labourers who carried out the NOSHEB Development Plan. Today, a new generation of hydro boys and girls is working in the Highlands and hoping to see hydro power come into its own as one of Scotland's sources of renewable energy.

Heroes abound in this story: maybe that's why I have found it such a compelling tale. Tastes in heroism are always changing, but as the twenty-first century starts and we begin to realise just how seriously our environment has been damaged by industrial development, hydro power, as a proven form of clean, sustainable energy generation is a highly valuable national asset: a real Scottish hero.

But hydro power began in another age, one when environmental considerations were chiefly concerned with preserving landscapes (and land values). This book traces the development of hydro power in the Highlands from its private and public beginnings to the rise and fall of the NOSHEB. It concludes by examining hydro power's role in today's privatised electricity supply industry.

Along the way, the reader will meet some of the story's personalities. While dams and power stations stand inscrutable and pristine as if they had just been delivered, the reality of hydro-electric development is very different. The construction of hydro-electric schemes makes a lot of noise and mess, and involves and affects the lives of many people. In this account of hydro power in the

Highlands, I have told the story of some of these people and the part that the hydro-electric adventure played in their lives. This book is a journey through time, and across and beneath the Highland landscape. Above all, it is not just a story of technology and politics but of people. We see them in NOSHEB tunnels choked with gellie-reek, ranging the Scottish landscape in search of possible hydro sites or peering anxiously from Highland homes at the doings of the NOSHEB's contractors. I was lucky enough to be able to talk to some of these people and the resulting material gives the reader a chance to hear some of the Highlands' hydro story at first hand.

Hydro power is not always a wholly benign system of power generation. All over the developing world, dam construction has made massive social dislocation the price of modernisation policies that are dictated by urban development priorities. The construction of the Kariba dam on the Zambezi valley in the late 1950s caused the displacement of the Tonga people, a 60,000 strong tribal community. The reservoir flooded the fertile river banks and the Tonga were dispersed and resettled on poor quality land, with their social cohesion shattered. On a much smaller scale, hydro development in the Highlands can also be disruptive, as we learn from Iain Mackay, once of Pait, as he describes what it felt like to have your home submerged in the name of progress.

Hydro power, at present, generates around 15% of the world's electricity so we must acknowledge its limitations. But this form of power has crucial advantages: it is renewable and sustainable. If the developed world, with its motorways, hospitals and global shopping malls, fails to maximise its use of such renewable energy sources then that very world cannot be sure to survive. Hydro-electric power offers to make a significant contribution to a sustainable future: the only future there is.

CHAPTER I

Who's Who in Hydro-electricity

How great the change since then [the Glencoe Massacre]! Though hardly yet a suggestion of what will be when the Highlands develop their natural industries through water power [and] beat the landlords and the scenic sentimentalists. There will never again be a repetition of the defeatism of the Clearances. The folk will come into their own. God hurry the merry day!

Neil Gunn, *Off in a Boat*, 1938

THE TITLE OF THIS BOOK, *The Hydro Boys*, is imperfect shorthand for that mixed bunch of enthusiasts who have dreamt of and worked for the development of hydro-electric power generation in Scotland ever since the end of the nineteenth century. The name was first used by Highlanders in the 1940s and 50s to refer to the men employed by the contractors carrying out the North of Scotland Hydro-Electric Board's Development Plan. In this book's title it means all those who played a part in the great Highland hydro power adventure. They came from all parts of society: military men, engineers, major and minor entrepreneurs, landowners, financiers, civil servants and politicians along with navvies and their modern counterparts. But despite their differences, they shared the same enthusiasm for the adventure of obtaining power from the (almost) natural flow of water for the benefit of society and also, of course, for themselves. For many of these people the hydro-electric project was a crusade with a marvellous goal: the prize of affordable power for all from Scottish rainfall. There was also the prize of financial reward. The setting up of a utility like hydro-electricity offers investors the chance of considerable long-term gain. Hydro construction also offered great opportunities to the Highland labourer. The Development Plan royally rewarded those workers who had enough skill, bravery, endurance and luck to make the most of its challenges.

The fuel needed to generate hydro-electricity is genuinely free at the point of delivery but the apparatus needed for such power generation is very costly. The design and construction of dams, aqueducts (pipes carrying water above and below ground) and power stations calls for mastery of engineering, geology, planning regulations and economics as well as access to lots of venture capital, public or private. Inevitably then, the prime movers of hydro-electric development are financiers and politicians, city men in suits with eyes firmly on the balance sheet. But these VIPs depend utterly in their hydro-electric endeavours on the knowledge, creativity and multifarious skills of civil engineers in whose ranks can be found some magnificent hydro boys. Mechanical and electrical engineers also have a key part in hydro design. The electrical engineers, Merz and MacLellan, took a leading role in promoting hydro-electric development in the 1920s.

However, it is civil engineers that have been most associated with campaigns for the promotion of hydro power. Sir Alexander Gibb's career at the start of the twentieth century is a prime example of the profitable links that can be forged between the civil engineering profession and political decision-makers. Tom Johnston knew just how much James Williamson's civil engineering expertise would influence the wartime deliberations of the Cooper Committee that eventually recommended the creation of the North of Scotland Hydro-Electric Board.

The professional title of Civil Engineer only began to be used in the first quarter of the nineteenth century. 'Civil' here describes the essentially non-military nature of the job. Yet the work done by this group of earth-shifting, bridge-building, tunnel-boring, world-shaping individuals and the workers they control and direct, does have its origins in the construction skills developed by military engineers.

In 1820, the great Scottish engineer Thomas Telford was elected first president of the Institution of Civil Engineers (ICE). A member, Thomas Tredgold, outlined the function of civil engineers at a meeting of the Institution in the same year. He declared: 'Civil engineering is the art of directing the great sources of power in nature for the use and convenience of man.' The civil engineer's job is to provide society in peacetime with the physical framework it needs for its survival. By the early nineteenth century, the massive changes wrought in Britain by the Industrial Revolution meant that society's needs had increased in scale and sophistication far beyond roads and bridges. The populations of the new industrial towns faced unprecedented problems of water supply and sewage disposal. Industrialists needed factories for their manufacturing projects and the cheapest possible housing for their workers. Railways, canals and shipping were vital for accessing the new economy's growing global markets and the new capitalists relied on civil engineering expertise in the construction of every bridge, terminus, dock and warehouse in their trading networks. At the end of the century, another industrial revolution occurred. Eventually, the public supply of electricity would completely transform life in the industrialised countries for almost everyone, at home and at work. Civil engineers were at the forefront of this revolution, taking responsibility for power station design and for supervising the construction and management of electricity generating operations.

Maybe it was the sewers, but civil engineers, with their weather stained mackintoshes, never achieved quite the same social standing as architects. This is odd when you consider that the work of any modern architect assumes a pre-existing infrastructure created by civil engineers. Not much good having a Divine Architect if you haven't got a Divine Civil Engineer as well. Society depends on the infallible undertakings of civil engineers simply to operate from day to day. Roads, sewers, airports, harbours and all the rest are not there because the residents of the developed world are the chosen beneficiaries of nature's bounty. Rather, our fortunate world is the way it is because people have worked out exactly how to make it that way and then organised its construction for us. Society's structural necessities are the end product of the meticulous diligence of dedicated but largely unsung professionals. The work of civil engineers is like housework: largely unremarked unless it is forgotten or badly done.

In their day, men like Telford, Rennie and Brunel were the celebrated stars of the society they served: their achievements were new and miraculous. In the twenty-first century, civil engineers are anonymous and their proficiency is taken for granted as the memories of their contribution to post-war reconstruction fades. Yet our lives depend more than ever on their expertise.

The early fabricators who created the bare bones of the pre-industrial infrastructure had become, by 1828, civil engineers to the new, busy society spawned by the Industrial Revolution. By the end of that century, electricity was beginning to be generated and consumed across the industrialised economies. Civil engineers were to be found at the forefront of one particular new technology: designing and constructing hydro-electric power generating schemes.

Just as hydro power promoters depend on the skills of civil engineers, the success of the plans drawn up by those engineers rely on the skill and stamina of their labourers. The Highland climate and terrain certainly posed extreme challenges for the workforce involved in the execution of the NOSHEB Development Plan.

Unlike England and southern Scotland, the Highlands saw relatively little of the civil engineering activity which accompanied industrialisation during the nineteenth century. When the navvies did arrive, they seemed to the Highlanders to be a very uncivilised race, despite being the harbingers of ultra-modern developments like canals and railways. In 1803, the Highland population had its first sight of major civil engineering activity. The cutting of the Caledonian Canal connected existing lochs to make a waterway between the east and west coasts of the Highlands and took nearly twenty years to complete. Men came from all over Scotland and Ireland to work on the canal which was designed by Thomas Telford. Highlanders joined the workforce and learnt the use of mattocks, picks and axes for the first time. The poet Robert Southey was deeply impressed by the spectacle of men, horses and machines at work on the Canal: 'such a mass of earth was being thrown up... that men appeared in the proportion of emnets to an ant-hill amid their own work.'

The next major civil engineering project undertaken in the Highlands was the extension of the Scottish rail network in the 1860s and 70s. The labour force came from the same mixed British and Irish origins as the canal cutters. So did the men who trekked to Foyers at the end of the century to start building the first commercial hydro power scheme in the Highlands. As we shall see, these men were often ignorant of the eventual purpose of their labours. But without the resilience and capabilities of their workforce, the plans of the engineers could not have been realised.

Workforce fatalities were routine on these early civil engineering jobs and would continue to be so until after the great hydro project in the north was complete. If society tends to overlook the skillful role of civil engineers, it knows even less of the travails and sacrifices of their workforce.

Hydro-electric power generation offers civil engineers the chance to shine as designers, technicians, environmentalists and economists. The first thing an expert in hydro-electric design has to do is to look at the landscape. Assessments of its geology, rainfall and run-off must be made before decisions can be taken about the optimum lay-out of the apparatus needed to generate electric power from rainfall. The resulting creation of dams, reservoirs and tunnels surely amount to a grown-up version of playing around with an existing water flow, like children in a burn have always done. But civil engineers designing hydro schemes get whole rivers, whole mountains and whole rainfall catchment areas to play with and they have thousands of men and machines at their command, dedicated to putting their plans into action. Hydro-electric designers walk up summits and along ridges and, looking down, decide how best to use the environment to make electric power from rainfall and gravity. Their skills and knowledge work together to turn on an incandescent light bulb from the depths of a rain cloud.

Financiers are not the only ones dependent on civil engineers for the execution of their hydro-electric plans. Politicians all over the world have turned into hydro boys, enthused by the potential for social change that they believe hydro power can bring. Tom Johnston, Labour stalwart and Scottish patriot, believed that hydro power could solve the problems of a whole region: the Scottish Highlands. His passionate belief in the North of Scotland Hydro-Electric Board's Development Plan allowed him to overlook the exclusion of trades unions from the Board's construction sites. For this hydro boy, the end justified the means. He infected other politicians with his enthusiasm, Emmanuel Shinwell, Aneurin Bevan and Jennie Lee among them. The last two were seen visiting the Affric-Beauly Scheme when it was under construction in the early 1950s. Bevan had his shirt off in the sunshine and the pair were unaccompanied. They were part of the hydro adventure, at least for the day.

It's probably fair to say that most of the Highland population also became hydro fans during the early days of the NOSHEB. The large number of locals employed on the construction of the Board's Development Plan earned more money than Highlanders had ever earned so near home. At the same time, increasing numbers of Highland residents were able to enjoy the luxury of electric power, courtesy of the Board's subsidised connection policy. When the Board was fighting for its political life in the 1960s, it received crucial and heartfelt support from its local customers. Thousands of remote dwellings had been connected to the National Grid, benefiting occupants who could never have afforded the true market price of connection.

However, public opinion was not always on the side of the hydro boys. In good adventure story tradition, the hydro boys' projects attracted dogged and vehement opponents. Hydro planners and promoters had to show the same granite determination when securing official sanction for their plans as their labourers displayed above and below ground in the watersheds of northern Scotland. The identities of those who wanted to wreck the hydro boys' projects varied over time. But they did share one dark characteristic. Those who routinely opposed hydro power did so for reasons of absolute self interest despite claiming to speak for the good of the entire nation. Hamish Mackinven, Assistant Information Officer with the NOSHEB (1952-84), described these enemies of hydro power development most succinctly: 'Private greed hiding behind public sentiment.'

The public sentiment most often exploited to justify attacks on hydro development was opposition to any economic development of the Highlands. This attitude was part of a longstanding, nationwide reaction to the effects of the Industrial Revolution on the British countryside. All over the lowlands of England, Scotland and Wales, factories and slum dwellings proliferated; successful entrepreneurs couldn't wait to quit the urban scenes of their manufacturing triumphs and build themselves country retreats. The rural idyll was supposed to guarantee the wealthy city dweller restorative peace, quiet and natural beauty. It must therefore be protected against intruders at all costs.

This split in British attitudes to town and country took on a violent intensity in perceptions of the Scottish Highlands. The dramatic

Highland scenery provoked much of this intensity. The region's mountains, lochs and jagged coastline offered the visitor a sense of aboriginal wildness: a perfect antidote to industry's utilitarian transformation of the lowland landscape. Moreover, because of the Clearances, much of the area was more or less devoid of people by 1850. Beauty and emptiness were two sides of the same coin to tourists from the crowded south.

Queen Victoria was personally responsible for deepening the intensity of British feelings about the Highlands. Figurehead of nation and empire, she was still a woman of her time, seeking emotional solace in the Scottish countryside, heedless of the difficulties people were actually experiencing there. In 1848, Queen Victoria and her husband visited Deeside for the first time and both were enchanted. The Highlands had been convulsed by the effects of famine since 1846: food riots were regularly covered in the national press. Yet the Queen was still able to write about Deeside in her diary: '...it did one good as one gazed around... [the place] seemed to breathe freedom and peace and to make one forget the world and its sad turmoils.'

Victoria's blinkered love affair with the north of Scotland encouraged hundreds of well-heeled imitators to buy Highland estates. Once ensconced, the men followed Albert's example and took to the hills in pursuit of game. This was more plentiful in the north than in the rest of Britain where industrialisation had degraded wildlife habitats. The royal couple enjoyed Deeside in different ways. The Queen meditated, while her Consort exterminated. Both activities shared one vital pre-condition: total privacy. Communing with nature is supposed to be a solitary business: apart from her own servants and carefully selected members of the Scottish aristocracy, the Queen preferred not to see anyone very much while off-duty in Scotland. Deer forests, where the game was reared and protected until the guns were ready, must be closed to any intruders who might threaten the livestock or get in the way when the hunt was on. Another entry from the Queen's diary emphasises the importance of keeping people out of the Highland picture: '...a great herd [of deer were] running down a good way, when most provokingly two men who were walking on the road – which they had no business to have done – suddenly came in sight... and the sport was spoiled.' This tendency to wish people out of the landscape was shared by all those who followed the Royal example and maintained property in the north.

Queen Victoria's passion for the Highlands nearly made a big difference to the eventual shape of the hydro-electric industry in the Highlands. Her first choice when property-hunting in the Highlands was Ardverickie, an estate on the south shores of Loch Laggan. Transformed into a reservoir, the loch became of crucial importance to commercial hydro power in Lochaber. Put off Ardverickie by the midges, the Queen chose Deeside instead, an area subsequently undisturbed by hydro-power development.

By 1900, this reverence for the 'unspoilt' state of the Highlands had become part of the national mentality. Intellectuals as well as ordinary working folk, who would never get closer to the Highlands than a charabanc excursion, felt the region was uniquely important to the nation. In the 1890s, the British Aluminium Company's pioneering hydro-electric development of the Falls of Foyers, Scotland's 'premier waterfall', was condemned by art critics, travel writers and landowners because it was going to 'spoil Scotland'. Subsequent attempts to exploit the hydro-electric potential of the Highlands provoked the same reaction: a refusal to welcome any changes which might threaten the Highlands' status as a pristine retreat from the industrialised lowlands of Britain. Scots from the lowlands, often descended from Highland stock, had a special reverence for the Highlands which led them to be deeply suspicious of Highland power generation schemes being set up for the financial benefit of southerners. As one commentator, Norrie Fraser, put it in the 1950s, 'To the people of Scotland, now concentrated in the Clyde Forth area amid the debris and dolour of mining and industrial development, climbing their grim cliff dwellings in the Scottish tenement, and having sacrificed the starry firmament on high for the spluttering street lamp, the Highlands of Scotland were a symbol of that inviolable beauty on which the dirty grasp of the industrialist must not fasten'. This intense protectiveness towards the Highlands played a powerful part in attacks, in and out of Parliament, on plans for hydro-electric development in the north.

Some of the most vehement opposition came from property

owners in the Highlands. Hydro development did have some supporters among the more responsible Highland lairds. Lord Airlie, the first Chairman of the NOSHEB, and the late Earl of Cromartie, for example, were firm believers in the need to modernise Highland living standards. Many landowners built small hydro schemes on their own estates: the lairds had no argument with the scientific principles involved. But the installation of large public supply schemes had very little to offer most Highland estate owners. Certainly, the lairds did not need electricity themselves. They had plenty of cheap labour to hew wood and draw water for them and by 1914 many estates had installed small generators to power their lights. Most landowners retreated to their more comfortable southern accommodation during the winter, so heating was not their major concern.

Moreover, there were plenty of reasons beyond their professed love of the Highland landscape for Highland landowners to oppose hydro-electric development. As we have established, their enjoyment of the Highlands depended on the uninterrupted and complete control of their properties. Allowing hydro engineers access to their lands and the right to interfere with the waterflow there threatened all that Highland landowners held dear. Not only would hydro-electric development threaten an estate's privacy and autonomy, but also seriously jeopardise the value of its sporting potential. Hydro power was assumed to be especially disastrous for salmon-fishing.

The idea that hydro-electric development would bring commercial prosperity to the Highlands actually made it even more unappealing to landowners. Their enjoyment of the Highlands depended on the absence of significant economic activity there. The lairds' Highland lifestyle required a supply of cheap labour that might disappear if widespread industrial enterprise pushed up wages in the north. A charming old lady from the Black Isle told me about her landed parents' hatred of all things hydro. Her mother, though born in England, had refused to entertain the idea of any change at all being inflicted on the Highlands. She had been especially horrified by the prospect of 'gangs of Irish labourers' at large in the north. The family was happy using candles and doing without central heating and simply did not consider the difficulties faced by the rest of the Highland population.

The political power of the British aristocracy underwent a general decline in the last half of the nineteenth century, as democratic reform and economic developments undermined its land-based political strength. How then were the Scottish landowners able to exert such a forceful influence against hydro-electric development? One possible answer is the absence in the far north of a significantly large middle-class eager to embrace economic opportunities. The remote and inhospitable terrain and the extreme disruption to civil society caused by the Clearances combined to limit the growth of a middle class which might have been able to counter the influence of the estate owners.

Ironically, the one entrepreneurial group which did emerge in the north before the end of the nineteenth century was irrevocably allied to the landowners' determination to resist the intrusion and change threatened by hydro power development. Well before 1900, tourists were regularly following the Royal route northwards. Accommodating these visitors had become big business by the 1930s and those profiting from this traffic regarded the Highland landscape as a prime business asset – if it were to be damaged or degraded, then profits from the tourist trade would be seriously jeopardised. In the late 1940s, the hoteliers of Pitlochry refused to accommodate the NOSHEB consultants who had gathered there to collect information for the promotion of the Board's Tummel-Garry Scheme. The hoteliers were terrified that once the proposed 'alien' dams and reservoirs were in place, the Highland landscape would no longer offer southern visitors solace and delight.

Landowners tended to feel threatened by the Board and all its works. Hydro veteran James Black was pelted with stones by a well-known Skye landowner while working on an NOSHEB construction site in Glenmoriston in the 1950s. Nonetheless, the NOSHEB Development Plan was carried out virtually unscathed by attacks from opponents until the publication of the Mackenzie Report in 1962. In fact, by the end of the Second World War, most Highland landowners had resigned themselves to accepting change for the good of the region. Even opponents of hydro-electric development, like Lord Lovat, do not seem to have had too much difficulty in accepting compensation paid by the Board to riparian owners when it became due to them.

Some of hydro-electricity's fiercest opponents came from other sectors of the power generating industry. Their fundamentally self-interested motivation was often cloaked by claims about long term benefits to the British economy. Until its disastrous run-in with Margaret Thatcher in the 1980s, the coal industry was a major political force in Britain and its representatives consistently sought to discredit and undermine the efforts of the hydro boys. Hydro development had to win Parliament's approval and it was at Westminster that representatives of the coal lobby attacked hydro development while claiming to have the nation's economic interests at heart. One of hydro power's great practical advantages is that it reduces the country's coal bill – thermal stations were mostly coal-fired until the coming of oil, gas and nuclear power mid-way through the last century. For the coal industry this notion of reduced demand was deeply threatening. Before nationalisation, the private coal companies had their representatives ready to attack hydro power proposals in Parliament on grounds of national energy policy, economics, environmental conservation and the superior efficiency of coal-fired over hydro power generation. After 1945, maverick Conservative MP, Gerald Nabarro, pursued a terrier-like campaign against the NOSHEB in Parliament and in the press on behalf of the nationalised coal industry.

By the 1960s, the Board was facing serious competition from relative newcomers to the energy supply industry: oil, nuclear power, and later, gas from the North Sea. Each claimed to be the cheap and abundant energy of the future. Such claims convinced the authors of the Mackenzie Report of 1962 which effectively closed down the NOSHEB Development Plan leaving over 60 hydro schemes unbuilt. Later, we will look more closely at how this decisive victory over the Board was won by the hydro boys' opponents in an unholy alliance of enraged landowners, academic economists, the South of Scotland Electricity Board and a gaggle of civil servants who preferred the technocratic tidiness of nuclear fission to the big adventure of hydro power. The Board's ambitions were ultimately defeated by powerful enemies, all hell-bent on their own individual agendas.

CHAPTER 2

What's What in Hydro-electricity

EACH SCHEME THAT GENERATES hydro-electric power is unique, designed in precise relationship to the terrain where it is going to operate. Schemes are also designed for specific electricity supply purposes: the generating requirements of metallurgical smelting operations, for example, differ totally from those of supplying electricity to domestic consumers. Hydro schemes designed to power smelters have a very high load factor while schemes designed to provide domestic consumers with electricity have a low load factor. A scheme's load factor is defined as the ratio between a scheme's capacity to produce current and the amount of current it actually produces.

But despite the differences that exist between individual hydro-electric schemes all over the world, the essential technology is the same, and basically very simple: water is passed through a turbine which then drives a dynamo to produce electric current. A stream of water is made to create a stream of electricity.

The word turbine was coined at the end of the nineteenth century and comes from the Latin word turbo meaning anything that spins or rotates. Modern engineers, working on design principles laid down in the pre-industrial world of mills powered by wind or water, devised the turbine, a motor in which a shaft is steadily rotated by the impact or reaction of a current of fluid on the blades of a wheel. Steam, gas and wind are used to drive turbines; at the centre of the hydro story are turbines driven by water.

Turbines are directly descended from the water wheels used by pre-industrial society to make mechanical power. Different types of wheel were developed over the years to utilise the different conditions of water supply and water velocity presented by different landscapes. Water turbine design is shaped by the same considerations; the main types, the impulse turbine and the reaction turbine, both had water wheels for prototypes.

The major influence of water mill technology on hydro-electric design is seen in hydro's extensive use of mill-related terminology. Penstocks are high-pressure tunnels, the tailrace is where water goes after passing through the turbines and a cofferdam is a temporary structure used to keep a normally submerged site dry to allow construction work to take place. All these terms originally described elements of water-mill design as did the important term, the head. In hydro power generation, the head is the height through which water falls to enter turbines in a hydro power station. The critical role played by head in hydro power generation is highlighted in the equation P=QxH, where P=power potential, Q=quantity of water and H=head.

The energy inherent in running water was probably first noticed during the irrigation operations on which the survival of many early agricultural societies depended. Later, irrigation raised water from its natural level in streams and rivers so that it could be directed to water crops. One of the oldest known devices for achieving this profitable reorganisation of nature is the Persian saquia, a wheel with buckets attached to its rim. The wheel was set on a horizontal axle, which was turned by animal-power: donkey or camel, for example. The wheel was partly submerged in the river. When the wheel turned round, its buckets emptied their water on the riverbank. Such primitive engines can still be seen at work in the Middle East.

When a saquia or similar apparatus is standing idle and unharnessed, the current of the river water will automatically turn the wheel in the opposite direction to its irrigation mode. A current of flowing water turning a wheel in the same way powered the earliest water mills. In gristmills, this water-powered wheel turned the heavy grindstones together to change grain into flour, mechanising one of the most important tasks of the pre-industrial age. The unsung genius who first exploited the water wheel's reversibility was a pioneering civil engineer who had rearranged nature expertly for the benefit of humankind. Antipater of Thessalonica celebrated this development about a century before the birth of Christ: 'Cease your work, ye maids who labour at the mill, for Ceres has commanded the water-nymphs to perform your task.'

The Romans were aware of water-powered technology but

they had no need to rely on it – throughout most of the Roman Empire, slaves ground corn by hand. It was the Saxons who introduced water mills to the British Isles. The first official reference to this development came in a charter issued by King Ethelbert of Kent in AD 762. The millwright who built and maintained water mills became an important figure in pre-industrial society. Over 5,000 water mills were listed in the Domesday Book of 1086 and water power, augmented by wind power, was the chief source of mechanical power until the invention of the steam engine in the eighteenth century.

The mechanical effectiveness of water mills was inevitably limited by the amount of water available, the head, the efficiency of the water wheel and of the gearing mechanisms which transmitted the wheel's power to the grinding apparatus. These limiting factors were beyond the powers of pre-modern technology to reduce in any significant way, so, by modern mechanical standards, water mills were not very efficient. But they made a vital contribution to the major economic activities of the day. As well as grinding corn, water power was used in the production of textiles, an important British export from the twelfth century onwards. Three centuries later, metal workers in Yorkshire were using water power to operate their blast-furnace bellows. Eventually water power was being used by iron manufacturers all over Britain for many important operations including hoisting and crushing ore, drilling gun barrels and drawing wire. Before the age of steam, there were so many water mills powering important jobs for important people that the navigation of rivers and the whole issue of water rights became a frequent and serious cause of dispute.

In the early phases of Britain's Industrial Revolution, water mills drove some of the new machines. Ultimately, however, it was steam that powered the great technical and economic innovations behind Britain's industrial success story. Scientific and entrepreneurial skills were responsible for the widespread adoption of steam power, as was the abundance of cheap coal in these islands. Labour costs were minimal for the coal bosses, with miners' employment conditions in the eighteenth century little better than those of feudal serfs. Coal was king and the industry assumed the political clout it would retain even under public ownership right up to the 1980s.

However, the essence of the modern capitalist economy turned out to be change, constant change. By the end of the nineteenth century a new industrial revolution was underway, one powered by electricity.

People had known about electricity for over two millennia before it became the marvellous new power of the twentieth century. Around 600 BC, Thales of Miletus noticed that if he rubbed a piece of amber with fur, the amber could pick up bits of straw and feathers. The word electricity comes from elektron, the ancient Greek word for amber. Electrical effects, like the frictional energy that Thales produced and of course lightning, interested many curious folk after Thales, but it wasn't until the eighteenth century that the discoveries were made that would lead to the Electric Age.

For electricity to perform useful work, a continuous stream of it is needed. The first person to find a way of obtaining this steady flow or current was the Italian, Alessandro Volta. He made the first electric battery in 1800. But Michael Faraday discovered what proved to be a more useful way of producing electricity in 1831. A decade earlier, Danish scientist Hans Oersted, had shown that an electric current could be made to produce a magnetic field. Faraday was the uneducated son of a blacksmith and the protégé of the eminent chemist, Sir Humphry Davy. The dynamo he invented reversed Oersted's experiment and used magnets to produce electric current. The electric current used for electric lighting, heating and machinery is produced by the operation of dynamos or generators as they are also known. A dynamo produces electric current by spinning, and then transforming that mechanical energy into electricity. The mechanical energy required to spin the dynamo can come from a variety of sources including engines fired by fossil fuels like coal and oil, or turbines powered by wind, gas, steam or water. The turbine's rotary movement is ideal for driving dynamos – a bicycle lamp is lit when the bicycle's spinning wheel activates a dynamo. Movement is made to create electricity; the same operation is at the heart of every hydro-electric generating plant.

Solar energy is also essential for hydro power production. The sun's energy evaporates water from the oceans and also creates the winds which transport the resulting water vapour to mountainous landmasses. When the vapour condenses, some of it falls as rain over high catchment areas. The rainwater runs off into rivers which flow downhill and hydro-electric engineers put plant comprising water turbines driving dynamos in the way of as much flowing water as possible. Water flow drives the turbines and the turbines turn the dynamos to produce electricity. This immediate activation of the turbines is the key to one of hydro power's great operational advantages for public supply generation. Hydro-electric power can be turned on and off almost as easily as a light switch, unlike power from thermal stations fuelled by oil, gas or coal. Thermal power stations can take hours to achieve generating capability from cold start-up conditions. This means hydro power is ideally suited for satisfying peak load demand, which occurs every breakfast- and tea-time on every day of the year and calls for a quick response from electricity generators.

Faraday's dynamo is only one part of the technology that generates hydro-electricity. Water turbines had first been produced at the end of the eighteenth century and the subsequent efforts of engineers world-wide secured major improvements in their efficiency. The 1860s and 70s saw crucial developments in the practical application of electricity which caused an upsurge of interest in its generation and transmission.

At the Munich Exhibition in 1882, scientists successfully transmitted electric current over a distance of one mile thus fulfilling an important precondition for the creation of public electricity supply schemes. One of these transmission demonstrations used current produced by a generator driven by the flow of the River Isar, making it the first public use of hydro-electrically generated power. France, Italy and the other Alpine countries of Europe would be responsible for much pioneering work in the field of hydro-electric power generation, being much better endowed with mountains than they are with coal.

At the start of the 1880s, the first hydro power scheme in the Britain was installed at Cragside near Rothbury in Northumberland. The link between the advent of hydro power and other pioneering advances in applied electrical technology was neatly demonstrated at Cragside. The house and its revolutionary power supply belonged to Sir William Armstrong, the

Newcastle engineer and industrialist. He pioneered hydraulics, built bridges and manufactured arms, amassing one of the biggest fortunes in Victorian Britain. He was a friend of Joseph Swan who invented the incandescent light bulb, powered by electric current. Swan's invention appeared at around the same time as Edison's light bulb in America. Edison and Swan combined their commercial interests and co-produced the 'Ediswan' light bulb, which some readers may remember. Cragside was Armstrong's opulent showcase of technical achievement. The hydro-powered light bulbs were miraculously better than smoky paraffin or smelly gas lighting and there was even an electrically powered dinner gong. Royalty came to Cragside to see the new technology at work but more importantly the appearance of the light bulb was a persuasive signal for the commercial organisation of public electricity supplies.

However, although small private schemes like Cragside proliferated from the 1880s onwards, the technical problems involved in long-distance transmission of electric current prevented major public supply schemes from using hydro-electricity. Hydro power is usually generated in remote, sparsely populated mountainous districts. Profitable, populous markets would remain out of reach until the 1920s when the difficulties of long-distance transmission had been solved. But the commercial use of hydro power started long before then. The first commercial hydro plant in Britain was built in Northern Ireland. A railway from Port Rush to Bushmills, powered by water from the River Bush, began operations in 1883. Four years later, the track was extended as far as the Giant's Causeway, making a total length of eight miles. In 1885, the Bessbrook and Newry Railway, also driven by hydro power, was opened and both railways ran until the 1950s. (Hydro power caught on throughout the island of Ireland, where civil engineering training soon included an emphasis on hydro-electric power generation. Hence the presence of Irish-trained engineers among the design and construction staff of the NOSHEB Development Plan, 1944-75.)

Metallurgical smelters were among the other early commercial users of hydro-electric power. They generated and consumed hydro power on the same site and so did not need to transmit power over any significant distances. But once the problem of long distance transmission of current was solved and the construction of the National Grid had been started, the stage was set for the development of public electricity supply schemes powered by hydro-electricity. The North of Scotland Hydro-Electric Board was the biggest and most important hydro-powered public supply venture ever undertaken in Scotland. To finish off this technical introduction to the history of hydro-electric power generation in the Highlands let's look briefly at the different types of generating scheme built by the Board.

Civil engineers base their choice of hydro design for exploiting the hydro-electric potential of a certain river basin or rainfall catchment area on the answers to a set of simple questions.

How much water is available?

Does that water flow constantly or does its volume fluctuate significantly on a daily, seasonal and yearly basis?

How will a scheme's necessary water storage be achieved? Is there a suitable site for a dam and the reservoir it creates?

What is the optimum height above the power station that a reservoir can be placed so as to achieve the maximum possible head?

What demand can reasonably be predicted for current supplied by the scheme and how is that demand likely to fluctuate?

Schemes designed according to the answers to these questions fall into four main types. In practice, most schemes combine elements from more than one of these types.

1. Run-of-river Scheme

The first we shall look at is the run-of-river scheme that has minimal water storage, if any. This lack of storage means that a run-of-river scheme's firm output (the minimum amount of current it is also to produce) is dictated by the river's minimum flow. A barrage or weir is built across the river which increases the volume of water passing through the turbines. Turbines are usually built into the weir or situated in a power station downstream. Power generation at Niagara Falls and at Victoria Falls is achieved by run-of-river schemes using the high head provided by the Falls. The Achanalt

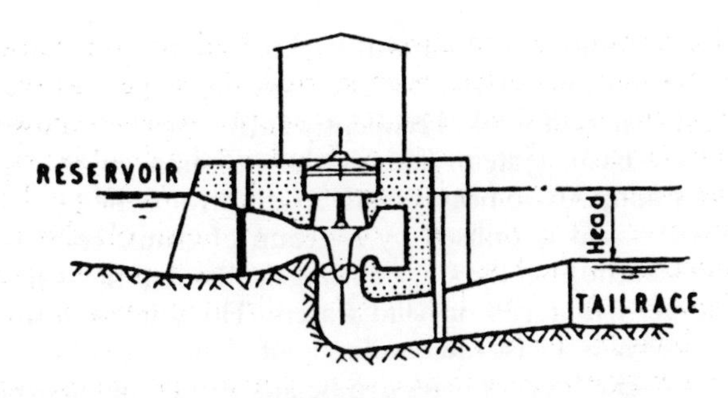

RUN OF RIVER

barrage and power station form a rather less dramatic run-of-river installation built by the NOSHEB as part of its Conon Valley Scheme.

2. *Dam Scheme*

A second type of hydro scheme has a dam to provide storage and to increase head. These schemes are usually situated in the lower reaches of a valley. Water supply on such schemes can be maximised by input from wide catchment areas. A valley's lower reaches

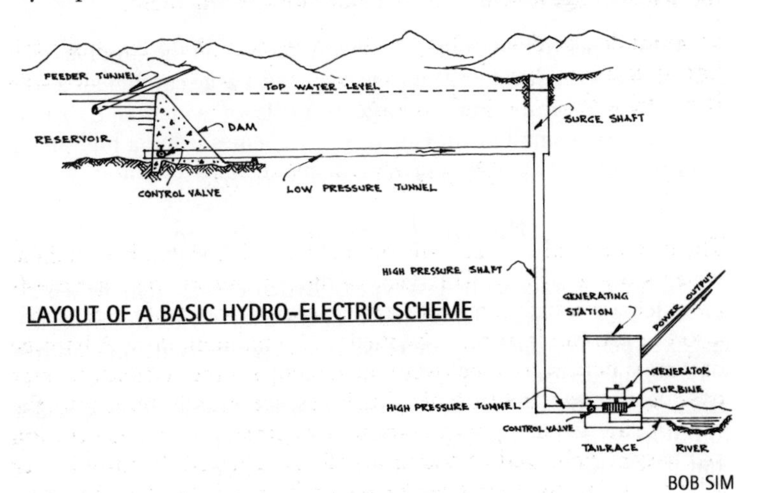

LAYOUT OF A BASIC HYDRO-ELECTRIC SCHEME

BOB SIM

often provide the sort of natural basins that are ideal for making into reservoirs thus saving the developer the time and trouble of major construction work. Frequently such schemes form part of a larger river basin system. This is the situation on the Tummel-Garry, Conon, Affric-Beauly and Strathfarrar Schemes, where a whole watershed is utilised by a group of dams, reservoirs and generating stations in the stage by stage control of the water stored in the upper tributaries or head waters. This sort of arrangement is also known as a cascade development. The generating potential of these lower valley schemes can be augmented by increasing the height of the dam although the expense involved in damming a broad river may not always be justifiable.

3. *High Head Scheme*

Another type of dam scheme maximises head by utilising the altitudes of high catchment areas or upland plateaux. In this sort of scheme water is delivered to the turbines from a reservoir, via tunnels and pipelines specially designed to withstand the very high internal pressures produced when water travels through a high head by the shortest, steepest route possible. On a scheme at Reisseck in the Austrian Alps, the head measures over a mile.

These types of hydro-electric scheme depend on the variable relationship between head and storage capacity. Large catchment areas can deliver correspondingly large run-offs of water and so can operate with a small head. Conversely, schemes using a high head can produce the same output with a smaller flow of water.

4. *Pumped Storage Scheme*

Pumped storage is the last type of hydro-electric scheme which we will look at here. These schemes do away with the need for large, expensive dams by increasing storage capacity in another way. Pumped storage schemes have two reservoirs, one above the other. Reversible turbine generators are used during off peak hours to pump water back to the upper reservoir from the lower one, so that it is ready to be used again for peak load power generation.

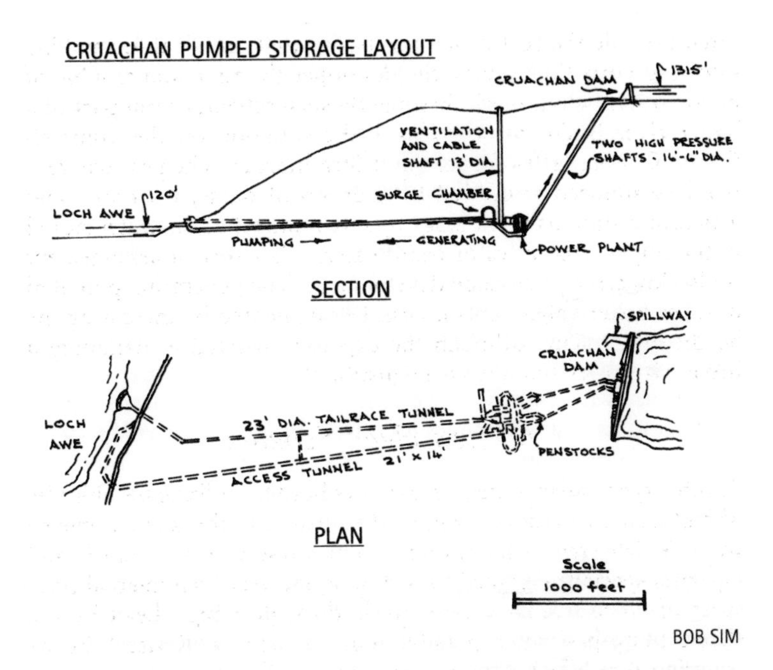

BOB SIM

The earliest recorded pumped storage installation in Scotland was a small private power station built in the 1920s at Walkerburn in the Borders. It was used to power a textile mill. Pumped storage generating schemes are ideal for locations where there is lack of sufficient catchment area to keep the reservoir topped up. They also require readily available off peak power from an external source to drive the pumps. The nuclear reactors at Hunterston and Torness and the coal-fired station at Longannet are the external sources of power for Cruachan, the NOSHEB's pioneering pumped storage hydro scheme, built in the 1960s.

The overall design of a particular scheme and the topography and geology of its location dictates the choice of dam type from a varied selection. A gravity dam relies on its sheer weight to resist the mass of water acting to overturn it. An arch dam supports the pressure of water by acting like an arched bridge laid on its side.

A buttress dam has buttresses like those you see in old churches. Rock- or earth- filled dams have a narrow concrete wall running across a flat valley with gently sloping banks of rocks or earth on each side of the wall. This sort of dam is rare in the Highlands where the glens are rocky, narrow and steep.

The ultimate factor influencing the designers of all hydro-electric schemes as they select from these options is the value of the estimated market for electricity. How much power is going to be needed, when and for how long? Answers to these questions determine the ultimate shape of the scheme. Economic considerations are paramount in dam design and all hydro-electric development must be able to meet strict cost and performance criteria. However, as the North of Scotland Hydro-Electric Board was to discover to its cost, economics is far from being an impartial science, based on objectively measured monetary quantities. Rather, it is an art controlled by subjective assumptions. In 1962, the economists who won over the Mackenzie Committee used such assumptions to discredit the NOSHEB and close down its Development Plan.

This brief survey of types of hydro scheme suggests a clear-cut categorisation not in fact found on the Highland schemes. As a paper published by the Institution of Civil Engineers put it, Scotland is a 'small country with limited water power resources and must turn every available stream to account'. To maximise water flow, many Highland schemes contain elements from the different types of scheme outlined above.

Highland topography offers hydro power developers a huge variety of opportunities and challenges. The north of Scotland may not have very high mountains or large rivers but its high annual rainfall and varied terrain compensate for these deficiencies. Hydro development is well suited to the Scottish climate with its cold, wet winters. The sight of black rain clouds massing over the Conon Basin in December can actually gladden the hearts of hydro engineers and put money in the bank for their employers.

This section owes a great deal to the excellent layperson's guide to hydro-electric power generation, *Power from Water* by J Guthrie Brown and T Paton (Leonard Hill, 1961), to *The Hydro* by Peter Payne (Aberdeen University Press, 1988) and to Bob Sim's generous scrutiny and his invaluable help with the diagrams.

CHAPTER 3

The Story of Hydro-electric Development in the Highlands

1890-1918: The Beginnings of Hydro Power in the North of Scotland

THE FIRST HYDRO-ELECTRIC adventurers in the north of Scotland were privately funded entrepreneurs. Many of them were landowners who set up small schemes to provide power for their own estates. The first electricity supply system in Scotland fuelled by water power to be extended to the public was turned on in 1890. This was the small scheme built by the Benedictine monks of Fort Augustus Abbey at the south western end of Loch Ness. The monks installed an 18 kW water turbine delivering power at 130 volts The water to drive the turbine was drawn from the River Tarff which the monks dammed with wooden beams set in concrete stanchions. A channel was built to take the water to a turbine that can still be seen today. A rake driven by a paddlewheel kept the water intake clear of debris. Power thus generated supplied the abbey and, later on, parts of the village as well. This system was used to operate the organ in the abbey's chapel and local legend has it that when the monks were playing the chapel organ, electric lights in the village houses dimmed. Like many early hydro-electric schemes, the one at Fort Augustus generated direct current.

The next significant hydro-electric public supply venture to be undertaken in the Highlands was, in respect of current, much more ahead of its time; it produced an alternating current like that generated by modern electric systems. The distribution voltage employed by this scheme was 415/240 which many years later became the National Standard. The man behind this apparently prescient scheme was, however, motivated in his hydro-electric undertakings by highly traditional concerns. Colonel Walter Blunt had married Sibell, Countess of Cromartie, in 1890. He changed his name to Blunt-Mackenzie and set about trying to restore the failing financial fortunes of the Cromartie Estate. To this end he was a keen supporter of the Spa resort at Strathpeffer. Blunt-Mackenzie was determined to use hydro power to enhance the facilities in Strathpeffer for the visitors who flocked from the south. He hoped that hydro-powered improvements would ultimately raise revenues for the estate. Many nearby estates like Ardross and Fairburn had been lit by water power since before 1900 and the Colonel was so keen to set up a public supply scheme using the same technology that he added some of his own money to that of the Cromartie Estate's investment in the scheme.

Issuing instructions from various military postings all over the world, the Colonel employed Deacon and Son from London to construct the scheme. They built a power station two miles from Strathpeffer, in the foothills of Ben Wyvis at Ravensrock. Here, a Pelton Wheel turbine driving an 80 kW set generated supply. To achieve the financial rewards of a larger distribution area, a cable was laid to extend the supply five miles from Strathpeffer to Dingwall. By 1903, only a little behind schedule, the system was operational. Its crowning glory was the splendid Gothic exterior of the Ben Wyvis Hotel lit up nightly by 420 separate electric lights.

This triumph was achieved in the face of considerable opposition, especially from the local gas suppliers who threatened the survival of the new power supply venture by slashing gas prices. Technological difficulties also dogged the Colonel's project. The first season's supply to Strathpeffer, a much-heralded event, suffered from frequent breaks in transmission. Costs spiralled and a significant demand for electricity failed to materialise. By 1909, the operation was being run by the Strathpeffer and Dingwall Electric Company Ltd which paid Blunt-Mackenzie £6,000 for the privilege of taking over the ailing enterprise. Most probably, this sum came nowhere near to covering the extent of his or the Estate's expenditure on the scheme. The cable from Strathpeffer to Dingwall had cost £2,000 alone.

The Dingwall Electric Co. was taken over in the 1920s by the Ross-shire Electric Company. The new owners closed down generating operations at Ravensrock and built a dam and power station on the southern shore of Loch Luichart, west of Strathpeffer in 1929. As part of its Development Plan, the NOSHEB refurbished the installation in 1954 for incorporation into its Conon Valley generation system. (Ravensrock was abandoned until the end of the century when a new landowner was able to take advantage of major changes in the energy market and use hydro power to generate electricity for sale to the National Grid.)

In 1910, the Duke of Atholl established a public supply scheme similar to Colonel Blunt-Mackenzie's. The Duke built a 130 kW power station on the Banvie Burn to supply electricity to Blair Castle and the village of Blair Atholl. Several other similarly small-scale, privately funded supply schemes were established in the north over the next two decades but their effective operational scope was limited by inadequate capital investment and their inability to maximise revenue by exporting current over significant distances. Ultimately, the most significant factor limiting all electrical development in the Highlands at this point was the low level of demand due to the general scarcity of disposable income in the region. Wage levels in the predominantly agricultural Highland economy were among the lowest in the British Isles at that time.

However, hydro-electric development in the Highlands was also beginning to take place in a commercial context which differed radically from these isolated and weakly-financed private supply schemes: aluminium production.

Aluminium is the most abundant metal in the earth's crust but it occurs there only as an oxide, Al_2O_3, known as alumina. The German chemist, Frederick Wohler, first achieved the difficult task of isolating the element in a pure form in 1827. By mid-century, the element, reduced from its oxide by complicated and expensive chemical processes, cost £20 per pound to produce. However, being light, strong and slow to corrode, the element's desirability was assured. In 1885, the newly discovered electrolytic method of producing aluminium brought its price down to four shillings per pound. This cost was further reduced by about 75% when, in France, PTL Heroult perfected an aluminium extraction process using an electric furnace. This process surpassed all others and made cheap electric power central to aluminium manufacture.

The Falls of Foyers on the south eastern shore of Loch Ness

were located in an area where the climate and topography provided the ideal setting for hydro-electric development. Accessible to bulk transport via the Caledonian Canal, the Falls provided perfect means for aluminium production. The Foyers Estate, which contained most of this catchment area, was bought by the British Aluminium Company which also acquired the British rights to the Heroult process. An Inverness-shire man, W Murray Morrison, was appointed as BAC's Chief Engineer.

Foyers, where the BAC began operations in 1896, was one of the earliest places in Europe where hydro-electricity was used in aluminium production. The company had some highly influential directors including, as scientific adviser, Lord Kelvin. Despite his reputation as a leading scientist, he had also been involved in a more traditional metallurgical venture, an attempt to manufacture gold from base metals. But the enterprise at Foyers relied on the very latest scientific developments.

The company had pre-empted serious legal challenges to its creation of a hydro-electrically powered factory by purchasing most of the catchment area on which its planned hydro-electric power generation depended. However, the project still attracted fierce criticism. While the Falls of Foyers are not the highest waterfalls in Scotland, their upper and lower sections are fed by the Foyers River and the lower section falls an impressive 100 feet into Loch Ness. The grandeur of the falls, especially with the river in full spate, had impressed visitors to the Highlands from the eighteenth century on, including Johnson, Burns, the Wordsworths and Southey. Christopher North of the influential *Blackwood's Magazine* claimed that the falls were 'worth walking a thousand miles to behold' and they were a regular attraction for the MacBrayne steamboat excursions on Loch Ness.

Opposition to interference with Scotland's premier waterfall was voiced, from the very heights of society, by what became, in effect, the first environmentally inspired protest against hydro-electric development. The anti-hydro agitators included representatives of the most exalted social groups: the Duke of Westminster, Canon Rawnsley, the art critic John Ruskin and the eminent travel guide writer, MJB Baddely. The opposition claimed that fumes from the smelter would destroy trees and plants for miles around the factory.

Baddely wrote to *The Times* asserting that the British Aluminium Company's scheme at Foyers 'is the greatest outrage on Nature perpetrated this century and the excuses made for it are inexcusable.'

The British Aluminium Company countered with what has since become the standard response of any industry accused of planning environmental vandalism, promising that the scheme would bring Foyers and the Highlands material benefits more significant than any damage it would cause. BAC's scheme did reduce the amount of water flowing down to the loch by taking water from the River Foyers just above the falls through a half-mile tunnel. This tunnel led to a set of Swiss-manufactured Girard turbines, which were still functioning when the North of Scotland Hydro-Electric Board took over the site in 1967.

In the intervening period, BAC's promise to bring prosperity to Foyers was amply justified. The initial workforce of 70 men grew to 250 and successive generations of locals continued in employment at the factory. The presence of a school, a shop and a church in this remote spot testified to the difference BAC had made. As a latter-day employee confirmed, 'The work at Foyers was extremely tough. But we were all lucky to get jobs. Without BAC we would have been really stuck.'

During the first few years of production at Foyers, the market for aluminium was slow to develop. However, by 1904, improved prospects for the product led the British Aluminium Company to begin work on a new hydro powered production scheme in Lochaber. This scheme used the water power available from the Blackwater River and the chain of lochs which stretch westward from Rannoch Moor to the coast at the head of Loch Leven. The Blackwater Dam was built, creating a storage reservoir of 24,000 million gallons of water. A conduit carried this water, along with more from diverted side streams, through six parallel welded steel pipes. These swept dramatically down the hillside to a factory-like power house beside the River Leven. The total installed capacity of the Pelton turbines was 23,725 kW, and with a load factor of over 70%, the plant's power output vastly exceeded that of any British hydro-electric plant then in existence. Even this huge capacity proved insufficient to meet the demand for aluminium created by World War I.

But the ordinary folk of the Highlands, people whose opinions weren't given much space in the newspapers or the Houses of Parliament, had reasons to feel uneasy about the hydro project on their doorstep. The labourers employed by BAC's contractors at Kinlochleven mostly came from outwith the Highlands. These rough outsiders had far more in common with the navvies who had laboured all over the rest of Britain on the industrial revolution's major civil engineering projects than with the local Lochaber population. The labourers' ragged appearance and wild behaviour combined with their itinerant lifestyle to alienate local society.

These pre-1914 developments at Kinlochleven are unique in that a vivid and highly articulate record of the workers' experience there in the first decade of the last century survives to inform us of the otherwise unimaginable conditions in which these men lived and worked. *Children of the Dead End* is Patrick MacGill's semi-autobiographical account of a navvy's life.

MacGill was born in 1891 in Donegal. Like hundreds of thousands of others, he had been forced by the extreme poverty of his homeland to travel to Scotland to find work. After potato-picking in Buteshire and railway work in Glasgow, MacGill 'padded it' to Kinlochleven. Here the British Aluminium Company used over 2,000 men to achieve the reordering of the Highland landscape required for its 'water-works'. The navvies themselves were wholly ignorant of the eventual purpose of their efforts: 'We turned the Highlands into a cinder heap and were as wise at the beginning as at the end of the task... All that we knew was that we had gutted whole mountains and hills in the operations.'

Many of the Kinlochleven workforce were Irish. There were also Scots from the Central Belt, Englishmen and a sprinkling of other nationalities. Their work consisted of moving earth and rock to create the British Aluminium Company's hydro-electric scheme at Kinlochleven with its dam, tunnels, aqueducts and pipelines. Their tools were picks, shovels, hammers, wheelbarrows and dynamite. Health and safety regulations were non-existent. The navvies' safety was not considered important and the contractors did not provide any medical facilities. Accidents were accepted as routine. The frequent fatalities often involved the dynamite used to blast away the rock. Blasting required holes to be drilled in the rock at a depth of about five feet into which the explosives could be inserted. One man held the steel drill or jumper in place on the rock to be bored, while five others took turns to strike the jumper's blunt end with a heavy hammer. In his usual unemotional tone, MacGill tells his reader that the men who held the drills were seldom wounded when an accident happened, death being the most usual outcome.

Fatally careless handling of dynamite was inevitable at Kinlochleven, given the heavy drinking which many men indulged in round the clock. MacGill's hero, Dermot Flynn, witnessed one such incident at first hand. Nettled by accusations that he was displaying unnecessary diligence, an Englishman, Bill, 'lifted his pick and drove it into the rock which we had blasted the day before. As he struck the ground there was a deadly roar; the pick whirled round, sprung upwards, twisted in the air like a windswept straw and entered Bill's throat just a finger's breadth below the Adam's apple.' Bill had struck an unexploded charge left from the previous day's blasting. Dermot and the rest of the work gang buried their workmate themselves but not before the immediate theft of the dead man's new boots reminds the reader of the unscrupulousness prevalent on the site.

At Kinlochleven the pay was sixpence an hour with limited overtime. Wages were paid daily which meant they were far more likely to fund the universal recreations of drinking and gambling than to be saved, even for a single day. Card schools ran continuously and there was a whisky shop on site. Disputes were rife and were often settled 'fistically' as MacGill put it. Crowds of onlookers would cheer the protagonists and gamble on the outcome. The accommodation, which the labourers had to pay for, was meagre and squalid. Men were crammed into windowless, ramshackle huts where they were provided with a bunk, a single blanket and the use of a hot-plate to cook food bought from the site store. The hut where Dermot Flynn lived at Kinlochleven was built over a spring and the floor was often underwater. Thieving was routine and the men had to sleep with their belongings under their pillows.

The navvies' freebooting lifestyle set them apart from Highland society. The men were, of course, not monsters but they may well have appeared so to Highland onlookers, the following anecdote evoking perfectly the sense of basic decency brutalised by brutal conditions. The money Flynn's hutmates had collected to send an invalid workmate back home to Skye was made redundant by the man's sudden death. There was brief talk of using the money to buy a cross for his grave but it wound up as a stake on the gambling table. Clancy, who eventually won it, was afterwards known as Clancy of the Cross.

Patrick MacGill's vivid stories of the navvies' life were so well received that he was able to take up writing professionally. He fought in the 1914-18 war and wrote about his experiences in France in the same vivid and unsentimental way he had chronicled his time at Kinlochleven. *The Great Push*, published in 1916, did not glorify war at all but celebrated the spirit of the men who had had to endure it. All his writings about the war had the same autobiographical inspiration as *Children of the Dead End* and they sold well.

However, after the war, his attempts to write fiction did not have anything like the same success. He married Margaret Gibbon, a novelist, and they emigrated to America. Neither, however, was able to find a way to make their literary talents pay in their new country. After a few years, they moved to California with their three daughters, hoping for employment as cinema scriptwriters but the Depression ended these hopes. MacGill continued to try fiction but none of his efforts were financially successful. His last years were spent in poverty and ill-health. The man who had shown the prosperous beneficiaries of the Edwardian Age the shocking details of life at the dead end of society died forgotten.

In 1909, when the Kinlochleven scheme was finished, it had the capacity to produce 7,500 tons of aluminium a year, twice as much as the amount consumed in the UK. The 1914-18 war radically reversed this over-capacity as aluminium was used for weapons and aircraft manufacture. The resulting new levels of demand forced the British Aluminium Company to expand. Balfour Beatty was employed to bring water from Loch Eilde Mhor into the Blackwater Reservoir. To accomplish this technically tricky task in wartime conditions, Balfour Beatty supplemented their labour force with troops deemed unfit for front-line duties and German prisoners of war

Demand for aluminium continued to grow after the hostilities were over. The war had been a good advertisement for the virtues of the versatile metal and by 1918, engineers and designers were much more aware of its vast practical potential. Only twenty years before, it had been considered semi-precious because of the great cost of producing it chemically. Now there seemed no limit to its uses.

The British Aluminium Company responded to this rise in demand with plans for a second Lochaber smelter, powered by water from Loch Treig and Loch Laggan. Parliament granted provisional approval but fierce local opposition forced the British Aluminium Company to cancel the project. Inverness County Council wanted Fort William to utilise the industrial potential of these lochs while locally there were strong objections to water being diverted from one catchment area to another. The British Aluminium Company drew up new plans for a power station and aluminium reduction plant at Fort William which satisfied both sets of objections. Water would be brought from the River Spey via a dam on Loch Laggan.

The most impressive part of this second British Aluminium Company scheme in the western Highlands was 15 miles of pressure tunnel driven through Ben Nevis, the longest tunnel in the world at the time. The massive scale and extreme complexity of the civil engineering required by this scheme combined with the economic uncertainties caused by the collapse of the post-war boom to delay its completion until December 1929. Up to 3,000 men were employed on the project but in conditions vastly improved since the days of Patrick MacGill. Perhaps the horrors of the Great War had made the men that survived them more compassionate, and certainly contractors now took more care of their workers. Alcohol was forbidden on site. The standard of accommodation provided by the contractors was hugely improved and included recreation areas, clothes-drying facilities, a canteen, stores and proper medical provision.

A further phase of the British Aluminium Company's Fort William project involved doubling the scheme's catchment area and increasing the head by raising the level of Loch Treig. The resulting increase in capacity meant the British Aluminium Company could meet the huge demand for aluminium created by the rearmament policies of the late 1930s. Further expansion took place during World War Two. This involved the diversion of the headwaters of the River Spey via a dam near Laggan Bridge and the installation of two extra turbines at the Fort William power station. These extensions were producing aluminium by 1943. The Fort William scheme had a catchment area of nearly 350 square miles and its annual output of hydro-electric power was the greatest in Britain at that time. At the heart of the British Aluminium Company's great technical and commercial success story in Lochaber lies a relationship of mutual benefit between hydro-electric development and the necessities of war. The west Highlands is not the only place where this relationship has flourished, as we shall see.

1918-1942: Successes and Failures

By the end of the 1914-18 war, commercial hydro power operations in Scotland were restricted to the activities of the British Aluminium Company. Hydro power had only a marginal role to play in the burgeoning electricity supply industry. But the British government was aware of the potential for water power development in Britain. In 1918, the Board of Trade set up a committee to investigate Britain's water power resources. The committee's chairman was Sir John Snell, a Cornish civil engineer with extensive experience of municipal and government service. In 1912, he had been a key witness in the arbitration leading to the state assuming control of the national telephone system. He had also been involved professionally and on behalf of the government in the growing public electricity supply industry. Snell was convinced of Britain's urgent need to modernise and the findings of the committee he chaired were accordingly forward-looking.

One effect of the Snell Report, published in 1920, was to make hydro-electric development more respectable. This new legitimacy increased the incentive for financiers and civil engineers to examine the economic rationale for exploiting Highland water power resources. As well as giving water power development the unambiguous stamp of government approval, the Snell Report also made specific practical recommendations which influenced attitudes to subsequent hydro-electric developments all over Scotland. The report stressed the wisdom of developing the hydro-electric potential of individual catchment areas as discrete units rather than diverting water from one catchment area to another. Snell also directed future developers to ensure that their hydro-electric schemes should directly and significantly benefit the inhabitants of the catchment area being used. Inter-war developers were not heedless of this direction but it was the North of Scotland Hydro-Electric Board which would take this principle to its ultimate conclusion.

By the 1920s, Britain trailed behind most of the world's other industrialised countries in terms of electricity consumption.

Consumption of electricity
(per head per annum in kilowatt hours) in the 1920s

Great Britain	110 kWh
California	1,200 kWh
USA as a whole	900 kWh
Norway	500 kWh
Canada	500 kWh
Shanghai	145 kWh

At a time when economies all over the world were reaping the massive benefits of the electrical revolution, the British Conservative government knew that action must be taken to promote greater national efficiency through increased use of electricity. The first step must be to improve provision for the public supply of electric power.

By the mid-1920s, the electricity supply industry consisted of a patchwork of independent generators of electric power, some privately owned and some controlled by municipal authorities. There were variations in size among these enterprises but, most critically, they were each generating electric current according to their own individual technical specifications. Most early generating schemes produced direct current but not all; some generated single phase current, some two-phase and some three-phase. Frequencies varied from 25 Hz to 83 Hz while voltages could be anywhere between 48 and 500. The main result of such chaos was to make everything connected with electricity very expensive. All appliances, even light bulbs, could not be manufactured to a consistent specification. Without the benefit of economies of scale, production costs for these items remained high. Gas had been in general use for lighting since the 1850s. Mantles for gas lamps were introduced for sale in 1885 and made gas light more serviceable and effective. This industry was able to make its uniform products cheaply enough to be affordable by most people.

Electricity therefore was used only by the rich. The small scale of many of the electricity generating operations meant that the cost of the current they produced was high. However, the suppliers could not make their product, the provision of electric power, cheaper by expanding their markets because of the lack of any technical conformity amongst them. Long-range transmission, vital for a profitable export of bulk power supplies, requires alternating current, which had not been universally adopted by 1920.

In 1925, Stanley Baldwin's Conservative government was well aware of the disadvantaged state of all aspects of the British electrical industry. What was needed was standardisation, interconnection and greater efficiency in current generation. These changes were absolutely essential if Britain was to compete internationally in the new Electric Age. To address these crucial problems, the government set up a committee headed by the leading Glasgow industrialist, Lord Weir. Lord William Douglas Weir (1877-1959) had taken his father's engineering firm to staggering success and extensive international contracts in World War 1. He had also made significant contributions in the sphere of wartime state planning. Employed by the Ministry of Munitions, he worked successfully to secure wartime supplies for the west of Scotland. With his high level of business and public service expertise, he knew the economic survival of Britain depended on the successful electrification of her economy. So it came about that a Conservative-appointed committee led by Weir, a well-known right-winger, eventually recommended the setting up of an electricity supply industry, financed, controlled, supervised and regulated by the state. Weir and his committee concluded in record time (their report took less than six months to prepare and contained persuasive evidence like the table included above) that such nationalisation was the only practical way of correcting Britain's technological deficit.

These unlikely nationalisers issued practical recommendations about how best to initiate improvements required by the electricity supply industry. To be effective, such recommendations had to be made statutory for the whole industry. Standardisation of current, voltage and frequency was a vital precondition to the construction of the National Grid, a network of high voltage transmission lines. The committee proposed the organisation of a Central Electricity Generating Board which would supervise the Grid and ensure that it would make electricity cheaper and more widely available. The Board's Electricity Commissioners would purchase power from generators and sell it back to distributors. The commissioners and the Central Electricity Generating Board were in charge of maintaining standards of efficiency in power stations. They also supervised the creation of the National Grid which developed piecemeal until its completion in 1945.

However politically surprising its centralist conclusions, the Weir Committee's achievements, in particular the setting up of a Grid capable of sending bulk supplies of current over many miles regardless of terrain, were extremely fortuitous for prospects of hydro-electric development in the Highlands. One particular element of Weir's innovatory regulations, the beginning of the National Grid, made a huge difference to the economic potential of Highland hydro power. Hydro-electric power generation is normally undertaken in the countryside, often high above sea-level. In such remote, sparsely populated locations, the local market for electric current is bound to be limited. The national standardisation to alternating current allowed the creation of the long-distance power transmission lines which make up the National Grid. This meant that power produced hydro-electrically in the Highlands could be efficiently exported to southern Scotland. Populous and heavily industrialised, the Central Belt offered hydro power developers profitable markets. Only the ability to access such markets made the huge capital investment involved in hydro-electric power generation commercially viable.

The importance of the hydro power option and its persistently controversial nature became abundantly clear in the years between 1926 – the year the Weir Committee's recommendations became law – and 1943 when Parliamentary legislation established the

North of Scotland Hydro-Electric Board. This period saw constant, public battles between hydro boys keen to exploit Highland hydro power and those who, for a shifting variety of reasons, tried to stop them.

Two out of the three large-scale Scottish hydro-electric schemes which were built in this period relied on long-range transmission of current to ensure their overall profitability but only one, the Grampian Electric Company, was based in the Highlands. However, all three schemes are noteworthy, if only because the promoters of each one had to overcome organised and vehemently expressed opposition before Parliament sanctioned their plans. The two non-Highland schemes are also important because they introduce us to two of the most influential hydro boys in the history of Highland hydro power and in particular in that chapter of it which belongs to the North of Scotland Hydro-Electric Board and its epoch-making Development Plan.

After the passing of the Central Electricity Act, there was a growing realisation of the economic potential of water power in the public supply of electricity. But promoters of all hydro-electric schemes still had to secure the approval of Parliament for any plans to exploit the water power of a particular area.

Founded in 1901, the Clyde Valley Electric Company was, by the 1920s, the largest of the independent Scottish companies supplying electricity from coal-fired power stations to rural areas and small towns. Hydro-electric development of the Falls of Clyde had been proposed as early as 1909 but without a sufficiently substantial market for generated power, the project failed to materialise. The Power and Traction Finance Company promoted the scheme in the 1920s having found a secure market by entering into partnership with the Clyde Valley Electric Company. Edward MacColl, Chief Technical Engineer with the CVEC, investigated the scheme's potential. He also looked closely into the associated issues of water rights and the safeguarding of amenities. He was able to recommend the construction of a run-of-river scheme. This diverted the natural flow of the Clyde through tunnels to two power stations, Stonebyres and Bonnington. The scheme was derided by the coal lobby who dubbed it 'MacColl's Folly'.

In fact, Edward MacColl had taken very effective steps to forestall criticism from the amenity lobby. He had avoided the controversy associated with reservoir storage by establishing that the Clyde's natural flow was great enough to obviate the need for storage. He made two further provisions to satisfy amenity-based objections. He ensured that high levels of compensation water would maintain the appearance of the Falls of Clyde and he paid particular attention to ensuring the appearance of the two power stations would blend harmoniously with the surrounding countryside. The whole scheme was finished in 1927 and the company's new hydro-electric operations combined with its existing thermal output in a highly effective way.

After 1926, hydro power development in Galloway could take decisive advantage of the introduction of the long-range transmission offered by the National Grid. Commercial success would be achieved by exporting current to populous markets.

The original report on the Galloway development had been made by Merz and McLellan, the consulting electrical engineers. They commissioned Sir Alexander Gibb and Partners to investigate the potential of a scheme centred on Loch Doon and the valley of the Galloway Dee. In order to obtain Parliamentary approval, the promoters had to defuse concerns that hydro power would damage salmon fishing on the River Dee and possibly inundate the ruins of Doon Castle.

Such was the attention paid in the promoters' plans to the survival of the salmon fishery by the installation of fish passes, that an engineer commented at the time, 'one might almost have been led to believe that it was a scheme for the preservation and improvement of the salmon fisheries on the River Dee.' Ingenious methods were used to make sure salmon were not discouraged from going about their natural business of returning to the river of their birth to spawn. Fish passes were installed at Loch Doon to help salmon negotiate the increased height of the new reservoir. The lay-out of hydro schemes was designed and adapted to safeguard salmon spawning grounds on the river bed and baffles were used to minimise water turbulence.

The great importance attached by Parliament to piscatorial and environmental considerations was made clear in the Bill which eventually granted Royal Assent to the Galloway scheme, passed in May 1929. It contained a clause which left no doubt about the high priority which would be have to be given to environmental and fishing interests in this and all subsequent hydro-electric planning decisions. 'In the construction of the works all reasonable regard shall be had to the preservation... for the public and for private owners, of the beauty of the scenery of the districts in which the said works are situated.'

To ensure that this 'regard' was maintained, the Secretary of State for Scotland was empowered to appoint a committee to make 'such recommendations as [it] thinks reasonable and proper for the preservation of the beauty of the scenery'. Amenity committees with functions identical to the one established for the Galloway development would henceforth be a compulsory feature of all hydro-electric developments.

Once the Act was passed in May 1929, the Galloway Water Power Company came into being. The promoters' success in Parliament had been helped by the company's promise to provide work for over 4,000 men on the project, at a time when the south-west of Scotland was experiencing severe unemployment.

The scheme, after a nerve-racking delay caused by uncertainty about the Central Electricity Generating Board's intentions to buy bulk supplies of current, was completed in October 1936. The gradual descent of the water from 700 feet at Loch Doon to sea level over a distance of 40 miles necessitated a cascade development and the construction of nine dams. The designer in chief for the whole project was James Williamson, then employed by Sir Alexander Gibb and Partners. Thanks to the newly created Grid, power generated by the Galloway Scheme could be sent north to the Scottish central belt and south to north-west England.

The third of the hydro-electric public supply schemes which overcame the obstacles of local opposition and Parliamentary regulation was promoted by the Grampian Electric Supply Company. In the wake of the Snell Committee's encouraging report, Parliament passed the Grampian Electric Supply Company Act in 1922. The Bill's successful passage in the House of Commons also owed much to its influential supporters who included the Duke of Atholl and the chairman of Lloyds Bank. The Grampian Electric Supply Company benefited from the financial support and technical expertise of George Balfour of Balfour Beatty, the Portsmouth-born son of a Dundonian naval worker. In line with Snell's recommendations, the firm's development plans for the southern Highlands exploited only a single catchment area, and the importance of serving the needs of the only local population had not been forgotten.

The export of bulk supplies beyond the company's immediate operational area in Perthshire was essential for its profitability. By 1927, the company made arrangements with the Central Electricity Generating Board concerning the purchase and transmission of bulk supplies of current with the Central Electricity Generating Board. Construction started in the following year on what would become, in its day, Britain's largest hydro-electric public supply scheme. The scheme raised the levels of Loch Rannoch and Loch Ericht and used dams, pipes and tunnels to maximise the catchment areas. Power stations built at Loch Rannoch and Tummel Bridge were operated in tandem to minimise the need for unsightly reservoir storage which had been expressly prohibited by the 1922 Act. In 1930, to meet increased demand from the Central Electricity Generating Board, Grampian Electric Supply Company raised the generating potential by building further tunnels and aqueducts to divert and capture extra water from neighbouring catchment areas.

By 1941, the Grampian Electricity Supply Company was serving a population of 414,000 in an area of over 10,000 square miles. Despite the relatively small return of 3.5% on its capital outlay, it stood as a fine example of what could be achieved by the exploitation of Highland water power. But the story of the inter-war period was more typically characterised by failed hydro-electric promotions: the three mentioned above were the exceptions to the general rule of defeat for the hydro boys. Even the successful schemes were forced to make major (and costly) accommodations with their opponents, for example the Galloway scheme's provisions for the local salmon fishery.

Despite this widespread opposition to water power development, hydro boys and their financiers saw opportunities for its exploitation all over the Highlands. Six major schemes were floated before 1943, despite the cost of promotion, which could be as high as £40,000. But when the Bills to approve these schemes reached Parliament, promoters failed to convince the House of a scheme's technical and economic rationale and their plans were defeated by well-organised and determined Parliamentary opposition. The Grampian

Electric Supply Company's scheme for hydro-electric development in Glen Affric, Inverness-shire, was rejected by Parliament in 1929 and a revised plan was thrown out in 1940. Four separate schemes for the exploitation of Loch Quoich and the adjoining catchment area north-west of Loch Ness were also defeated. Powerful Parliamentary alliances were dedicated to stopping the hydro boys at all costs.

The industrial objectives of the west Highland schemes promised to bring badly needed employment to the region, but this promise only antagonised Westminster MPs from other areas equally desperate for jobs. The Glen Affric development was also seen by some as a dire threat to that area's celebrated scenic beauty. Many English MPs, inspired by romantic visions of the Scottish Highlands as unspoilt wilderness enhanced by unique sporting opportunities, argued for the preservation of Glen Affric. Edward Keeling, Member for Twickenham, wound up his 1941 Commons attack on the Affric scheme by warning, 'It would create sinister changes in the landscape featuring huge white dams... wide stretches of rotting vegetation and slimy mud... and here and there the blackened skeletons of trees... projecting above the ooze'. Some Scottish MPs opposed the schemes, on the grounds of defending the Highlands from economic exploitation by outsiders or those unconcerned with the welfare of the native inhabitants. The emotive memory of the Highland Clearances intensified the special reverence felt by many Scots for the well-being of the Highlands.

Yet there was one strand of opposition which was not based on wholesale resistance to hydro-electric development but was inspired by the belief that such controversial exploitation of national resources should be undertaken by the state.

In 1941, Britain's perilous position in the war against Hitler was seized upon by some as another argument against major capital investment in projects not essential to the war effort. Yet one man did have the extraordinary vision to see, even at this most desperate moment of British history, the advantages to be gained by the systematic development of hydro-electric power in the Highlands. His was the vision of the ultimate hydro boy though his career as a politician never once gave rise to charges of adventuring or opportunism.

1943: Tom Johnston and the Founding of the North of Scotland Electricity Board

Tom Johnston's formative political allegiances do not suggest the sort of man Winston Churchill would one day enlist in his wartime coalition cabinet. Born in Kirkintilloch in 1881, Tom Johnston began his political career in local government there. Despite the comfort of his bourgeois background (his father had his own grocer's shop), Johnston's politics were socialist. He initiated numerous municipal enterprises in Kirkintilloch, including a community bank and a subsidised cinema. He then moved via the Independent Labour Party into national politics. In 1906, taking advantage of family connections in the printing business, he set up the weekly paper *Forward* as the official paper of the Independent Labour Party. He was to be its editor for the next 27 years. *Forward* became a class war comic for the proletariat. Issues were debated in its pages largely from an ILP perspective. The paper's anti-war stance led to *Forward* being banned briefly in February 1916 for allegedly treasonable activity.

Johnston also published two great historical accounts of the class wars that he believed had shaped Scottish history. *Our Scots Noble Families*, published in 1909, catalogued the crimes of the major Scottish landowning dynasties over the centuries. The damning irony of the title would have been relished by readers of *Forward*. In later years, his position in Churchill's coalition cabinet would bring Johnston into close working contact with various members of the very families he had attacked in *Our Scots Noble Families*. Published twelve years later, *The History of the Working Class in Scotland* was a more scholarly work. At its heart was a passionate condemnation of the miseries imposed by landlords on the common folk. So far, so very un-Churchillian.

In 1922, Johnston entered Parliament as MP for West Stirlingshire. Here began his association with James Maxton, David Kirkwood and the 'Red Clydesiders' group of left-wingers. By this time he had adopted the dress that he wore for the rest of his political career: a slightly old-fashioned, dark-coloured, double-breasted suit, hand-crafted polka dot bow tie, homburg hat,

Crombie overcoat and white silk scarf. The few times he was ever seen wearing anything else was when he put on deerstalker and tweeds to indulge in his hobby of fly-fishing.

He lost his Commons seat in 1924 but regained it in 1929 and the Labour Prime Minister, Ramsay MacDonald, promoted him to Junior Ministerial rank at the Scottish Office. He entered the cabinet as Lord Privy Seal, but his prospects of making a steady climb to a senior position in the Parliamentary Labour Party were scuppered by the turbulent events in Britain as global depression struck.

Despite his promotion under Ramsay MacDonald, Tom Johnston did not support the Labour Party's 1931 coalition with the Conservatives. Faced with the devastating effects of the world economic crisis, MacDonald and his Chancellor, Philip Snowden had responded with savage cuts in government spending. Under pressure from the banks, MacDonald resigned, entered into an agreement with the Conservatives and returned as Prime Minister of a national coalition in the General Election of 1931. The coalition had a majority of over 500 and the anti-MacDonald Labour Party was reduced to 46 seats. MacDonald was despised by many in the Labour movement and beyond for clinging on to the office of Prime Minister at any price, whilst blaming the unions for the crisis.

Johnston completely rejected MacDonald's analysis of economic events. For him, the cause of the catastrophe lay with the workings of international finance. He lost his seat but kept clear of the taint of MacDonald, whom Churchill derided as 'the boneless wonder'. Thomas Johnston regained his seat in 1935. His unwavering anti-appeasement stand must have impressed Churchill. Johnston also had a record of effective administration, including supervision of the St Kilda evacuation in 1930 and his work as Regional Commissioner for Scotland as Britain prepared for war in 1939. And there was another reason beyond the Scotsman's political integrity and practical efficiency that made Churchill want Johnston to become Secretary of State for Scotland in his wartime government.

In 1941, the greatest single concentration of shipyards in the world was on the Clyde. The need to guard against mutinous tendencies among the shipbuilding and munitions workers of Scotland was paramount for national security. The 1920s had seen 'Red Clydeside' convulsed by labour protests. Did Churchill need Johnston on his team to neutralise this potential threat of insurrection and to ensure that the vital Scottish contribution to the war effort did not falter? Most important for our story is the fact that Churchill was able to overcome Johnston's initial reluctance to enter the Cabinet by promising him decisive powers in planning Scottish post-war reconstruction. When Johnston suggested forming a Council of State for Scotland composed of former Scottish Secretaries of State from all parties, Churchill promised to support any measure which this body of luminaries unanimously recommended.

With this incentive to practical agreement in mind, the Council agreed to focus on the areas of agreement which already existed between its members. Development of water power in the North of Scotland was one such area. Whatever the reasons which inspired Churchill to give the old Independent Labour Party firebrand such an influential position, hydro-electric development could never have proceeded in the radical way it did without the indirect but fully committed support of the Prime Minister.

The first topic which Johnston asked the Council to consider was the potential for hydro-electric development in the north. An inquiry, headed by Lord Cooper, the Lord Justice Clerk, was set up in October 1941 to investigate the matter fully. Cooper's task was 'to consider (a) the practicability and desirability of further developments in the use of water power in Scotland for the generation of electricity and (b) by what type of authority or body such developments, if any, should be undertaken, and under what conditions, having due regard to the general interest of the local population and to considerations of amenity, and to report.'

The strength of Johnston's political position at this point was reflected in the personnel of the Cooper Committee. It included Neil Beaton, chairman of the Scottish Co-operative Wholesale Society and son of a Highland shepherd, John A Cameron of the Land Court, Lord Weir who participated despite his employment in vital wartime supply work, and James Williamson, civil engineer and expert in hydro-electric power generation.

Lord Cooper's attitude to hydro development was initially lukewarm; he believed that little could be added to the conclusions of the Snell Report over twenty years earlier. The committee, however, undertook a thorough examination of all previous evidence on the subject. It looked at the findings of Parliamentary Committees on Highland economic development, technical reports from representatives of the Electricity Commission and statements from local government and the Scottish Development Council. The metallurgical industry contributed to the debate via the representatives of BAC. Existing public electricity supply companies also gave evidence.

The opponents of hydro-electric development were included in the process. Scotland's fishery interests and the Association for the Preservation of Rural Scotland were also asked for their views. During the inquiry, the committee received all sorts of advice and instruction from the press, journalists being well aware of the passions that the issue of hydro-electric development could provoke.

The Committee's Report was published on 15 July 1942. It began by regretting the acrimonious and unproductive atmosphere in which hydro-electric development had been discussed since the Snell Report had been published: 'All major issues of policy, both national and local have tended to become completely submerged in the conflict of contending sectional interests.'

This clamour of conflicting interests allowed the Highlands' poverty and stagnation, as evidenced in heavy emigration, to continue unchecked. However, some positive signs of economic development in the Inverness and Easter Ross areas could be detected and failure to supply affordable power to these areas would jeopardise their further progress. The committee decided it would be too expensive to supply these districts with coal-fired electricity such a long and awkward distance from the coalfields. The obvious choice was to exploit the abundant fuel resources of the Highland watersheds. The total capacity of hydro-electric plant then installed in Scotland was no more than 315,000 kW. James Williamson knew very well how much greater this yield could be. His years of surveying the hydro power potential of the Highlands showed there were 102 practicable schemes which could be set up in the north. In total, these could produce another 450,000 kW.

The Cooper Committee's report shared Tom Johnston's vision of harnessing the great untapped potential energy. Just as importantly, the committee also backed his ideas about the sort of body that should manage this development.

Cooper's conclusions were translated into reality by the 1943

Hydro-Electric Development (Scotland) Act. This Act was passed without a division in both Houses, an outcome highly indicative of the widespread and influential support Johnston had secured for his hydro power project. The Act recommended the creation of a board authorised to manage hydro-electric power generation in the north, to be known as the North of Scotland Hydro-Electric Board.

The chief purpose of this new body was to reverse the bleak economic conditions prevailing in the Highlands. This was to be achieved in various ways. Industry would be encouraged into the Highlands by a supply of cheap power and an area-wide policy of affordable connections to the Grid for private consumers. These connections were to be made irrespective of real cost, and regardless of remoteness. Diesel generators and bottled gas supplied free of transport charges would be provided by the Board to customers awaiting connection. Crofters were encouraged to dry hay and grain electrically and the fishing industry was introduced to the benefits of electrically powered refrigeration.

All these socially-motivated policies were to be financed by exporting bulk supplies to the populous markets in the south of Scotland. This key aspect of Board policy was the result of the 1943 Act's famous Social Clause. Unprecedentedly radical for its time and accused by the Conservatives of being socialist, the Social Clause codified Johnston's key objectives for the Highlands in Subsection 3 of Section Two of the Act – that profits produced by the sale of bulk supplies of electricity be used for 'the economic development and social improvement of the North of Scotland'.

The idea of selling invisible energy across such vast distances and over such harsh terrain must have seemed as strange to folk then as dot.com manoeuvrings do to us now. Suffice to say that such a constructive way of linking up the poorer members of society to the latest technology could never have happened without the advent of the Scottish National Grid and, most importantly, Tom Johnston's political will to achieve such an egalitarian and positive outcome.

The Cooper Committee had also looked at a number of subsidiary but potentially controversial issues which were covered by the Act. These included extending the supply of electricity to the Islands, establishing rates assessment criteria for hydro-electric undertakings and experimenting with alternative fuels for the generation of power in the Highlands. The new Board would be able to purchase land or construct works without riparian owners being able to name their own compensation figures. The vexed amenity question was addressed with brisk decisiveness: the committee was sure 'that the complaints which have been made and the fears which are entertained on the score of injury to amenity have been seriously exaggerated'. The Committee also gave the extreme remoteness of potential hydro-electric generating locations as a good reason for not worrying too much about the look of things. After all, there was no-one there to see anything, only 'a handful of deer-stalkers, salmon anglers, gillies and gamekeepers and the adventurous spirits who have traversed the mountain districts on foot'.

From the Snell Report of 1922 onwards, Parliament had consistently stressed the importance of amenity and its preservation. But nobody was to be in any doubt of the future Board's robust justification for all aspects of its activities. The report ended self-confidently: 'But if as we hope and believe, the policy to which this report is a small contribution is to give the Highlands and the Highlanders a future as well as a past and to provide opportunity in the Highlands for initiative, independence and industry, then we consider a few localised interferences with natural beauties a small price to pay for the solid benefits which would be realised.' The NOSHEB inherited this confidence. It was soon to need plenty of it.

Born Fighting: The Board's Difficult Early Years

The Board owed its existence to a highly particular set of political circumstances: Johnston's commitment to 'doing something decisive for the Highlands' would almost certainly not have been so effectively realised without Churchill's support. But even such powerful backing did not mean that Johnston's project was given *carte blanche*. Parliamentary supervision of all hydro-electric development remained compulsory.

The Board which would put into practice Cooper's recommendations was to consist of five members. Four, including the

Chairman and Deputy Chairman, would be appointed jointly by the Secretary of State for Scotland and the Minister of Fuel and Power, and the fifth would be a nominee of the Central Electricity Board. The Board's first task, as laid down in the 1943 Act, was to conduct a general survey of water power resources in Scotland. Specific schemes were then to be selected for development and only if the Secretary of State were satisfied that there were no serious objections to a particular scheme would he sanction its execution. If any serious objections were made then the Secretary of State had to hold an Inquiry to address them before he could approve the scheme. As well as this check on the Board's actions, every construction scheme still had to gain Parliamentary approval. If either the Commons or the Lords objected to a particular scheme within forty days, the scheme could be annulled.

From the outset, the Board's activities were subject to serious scrutiny. Even before the war ended, its vulnerability to political attack was made starkly apparent. In 1943, Major Gwilym Lloyd George, Minister of Fuel and Power, headed a committee looking into the further rationalisation of electricity supply in Britain. It recommended the creation of a Central Generating Board in charge of fourteen regional distribution Boards, all accountable to it and thereby to the demands of national economic policy. Johnston was horrified by this threat to the North of Scotland Hydro-Electric Board's independence and to its mission to rescue the Highlands.

'Your proposals,' he wrote to Lloyd George, 'would drive a coach and four through the conception of the Act of 1943.' Johnston's political skills and influence ensured that this attempt to strangle the Board at birth was successfully resisted. However, with his departure from Parliament at the General Election of 1945, the NOSHEB lost crucial support in the Commons.

Lord Airlie, the Board's first chairman, attracted some particularly vicious criticism from opponents of the Board's plans who accused him of betraying his own class; his son had been forced out of the Perthshire Hunt in 1945 because of his father's leading rôle on the Board.

By 1948, the Board's scheme for Loch Duntelchaig by Loch Ness had been defeated by Inverness Burgh Council which wanted to reserve this loch for drinking water. The Joint Parliamentary

Commissioners, meeting in Edinburgh, supported the Council and forced the Board to abandon its plans. The Board's very first development at Loch Sloy attracted widespread criticism, some of it the result of the Board's poor presentation of its intentions. Dumbarton County Council wanted control of Loch Sloy for its own domestic and industrial purposes. The Board did win the battle for Sloy, by stressing the importance of its own national concerns over those of a single area and gained Parliamentary approval without the Board having to modify its plans significantly.

However, the next major project on the NOSHEB agenda was the expansion of the Grampian Electric Supply Company's scheme at Pitlochry. The influential Perthshire gentry united with local hoteliers in a campaign against more hydro-electric development there. The sensitive Lord Airlie had little stomach for the fight and, in 1947, conceded the chairmanship to Johnston, who held the position until 1959. His talents as a tough negotiator and his political contacts gave the Board an advantage which proved ultimately indispensable in enabling it to fulfil its founding intentions.

The NoSHEB and Highland Development

We had a great expectation of power all over the countryside... there are many things beside the production of electric power for industry. There is power for agriculture, forestry, fishery, tourist traffic, cheap transport and many other things. These are all bound up with the prosperity of the country.

Lord Airlie, House of Lords, 1943

We will be examining the realisation of the Board's plans in the next section but here is as good a place as any to assess Johnston's achievement in creating the North of Scotland Hydro-Electric Board. When he left the chairmanship of the Board in 1959 his aim of creating a strong economy in the Highlands had clearly not been achieved. However, the major target of extending the electricity supply throughout the Highlands and Islands was well underway and would be virtually complete by 1980. It is hard to imagine any other way this vital piece of modernisation would have occurred. But emigration continued apace, as Highlanders were forced to leave to find better employment opportunities and a higher standard of living. Why did Tom Johnston's plans for revitalising the Highlands economy through hydro-electric development come to nothing very much in the end?

There is, not surprisingly, more than one possible answer to this question. Hamish Mackinven is a great admirer of Tom Johnston and he told me what he thought was the chief reason for the Board's failure to achieve its economic goal for the Highlands. 'Even though the Act of 1943 and its Social Clause promised to give priority to stimulating the Highlands economically, when it came down to it, no-one was ever put in overall charge of doing that. Everybody at the top was an engineer and the Clause wound up looking as if it had been just a bit of PR.'

Another possible explanation of the Board's inability to bring prosperity to the Highlands can be found on the other side of the Atlantic Ocean with the Tennessee Valley Authority. The TVA was created in May 1933 as part of President FD Roosevelt's New Deal, an attempt to lift America out of depression through systematic state intervention. The TVA initiated a major programme to improve agriculture and attract industry to the valley. At the heart of this programme was an extensive plan for hydro-electric development. By 1953 the Authority had built twenty dams across the Tennessee River. In 1933, with only one dam in existence, the electricity-producing capacity of the area was less than 815,000 kW. Twenty years after the TVA's creation, this capacity had grown to 4 million kW. Flood control and the arrest of soil erosion in the area had also improved the region's agricultural prospects considerably.

In 1933, Tom Johnston was out of Parliament and grappling with the problems faced by Britain as the country tried, like America, to recover from the global economic crisis of 1929. The New Deal must have looked like a marvellously pro-active answer to the problems caused by world depression. A decade later, the TVA's hydro powered strategy seemed to offer a solution to the problems of the Highlands. The Tennessee Valley, by 1943, had started to catch up with American rates of industrial development, supporting Johnston's belief that cheap hydro-electric power could be the basis for industrial development dynamic enough to reverse population decline in the Highlands.

He wanted, above all, 'to get things done for the Highlands'. As Secretary of State, he rationalised and modernised the work of the Forestry Commission and fought hard to promote Scottish tourism, even sponsoring research into midge repellent. But to see the TVA as a model for economic transformation of the Highlands turns out to have been wishful thinking.

The TVA's operational area was vast compared to the Highlands. Johnston failed to see the importance of the major differences between the two areas. First, the Tennessee River and its valley are simply much bigger than any features in the Highland landscape. The river flows for 652 miles and the highest mountains in Tennessee are over 6,000 feet high – the implications of this difference in scale do not seem to have entered Johnston's calculations.

But the most important difference between the two areas is not one of physical but human geography. The TVA had been set up to improve the lot of the impoverished rural communities left behind in America's race for economic growth and the Tennessee population did benefit substantially from the TVA in terms of improved agricultural and industrial employment opportunities. In the Highlands the position was very different. By 1940, there was hardly anybody left. The area's rural poor had effectively been cleared from the landscape by the end of the nineteenth century. A vital precondition for local economic development was missing: people. There was no significant internal market for potential new industries. Moreover, the distance from external markets and from the suppliers of raw materials plus the area's inadequate infrastructure meant that using cheap energy as a stimulus for economic growth just didn't work. However hard Johnston and the Board wanted to stimulate the Highland economy, it may have simply been an unrealistic hope.

Ambitions to create a vibrant Highland economy had failed before. The commission administering lairds' estates, confiscated after the 1745 Jacobite Rebellion, set up linen manufacturing projects. But these enterprises failed to achieve any commercial success, let alone become the hub of further economic expansion as hoped for government in London. In 1803, the sound military reasons for starting to build the Caledonian Canal meant government finance was forthcoming. Government hoped this project would also provide employment for Highlanders and inspire the sort of economic transformation then being experienced throughout the rest of Britain. But when the canal was opened in 1822, emigration had not halted and Parliament was tired of complaining to the canal commissioners that most of the wages had been paid to 'natives of Ireland'. More recently, the history of the failed aluminium smelter at Invergordon and the uncertainties of the oil-rig fabrication industry on the Cromarty Firth seem to suggest that the Highland area is a particularly challenging environment for the creation of enduring industrial projects.

In failing to usher an Age of Industry into the Highlands, the NOSHEB had many honourable predecessors. Making connection to the Grid affordable for the majority of Highland residents was a truly progressive development for the region. If these connections had never been made, the Highlands today might still be the preserve of aristocratic sporting visitors, able to rely on private generators and poorly paid staff. The cost of supplying electricity to the more remote and inaccessible Highlands could only have been met by the inspired interventionism that created the North of Scotland Hydro-Electric Board. Johnston's role in that creation was undoubtedly epoch-making.

By 1960, over 80% of farms and crofts in the north of Scotland had been connected to the Grid. Many of these connections were highly uneconomic, but the Board continued extending its network through the 1970s and 80s. A diagram showing connections established by the Board in Aberdeenshire by 1960 shows a dense criss-cross of high voltage transmission lines where before there had been only a few single strands. This transformation, repeated all over the north, is surely Tom Johnston's finest memorial.

There are further ironies in Johnston's identification of the Tennessee Valley Authority as a socially benevolent development agency. By the mid-1950s, the TVA's programme of regional action had shed most of its social emphasis in the face of the big business imperatives. Full employment was achieved but not as much social justice as had been promised.

There is another, bleaker, paradox in the TVA story. Like the NOSHEB, the TVA owed its creation to a particular set of historical circumstances, namely, Roosevelt's New Deal. In the 1940s, the

TVA sought to ensure its profitable survival by finding a reliable customer for its abundant cheap energy supplies. Its hydro-electric generating capabilities convinced the American Atomic Energy Commission to choose Tennessee Valley as the location for the production of plutonium for its atomic weapons. The region had already some experience of munitions work with the production of explosives from locally mined nitrate compounds. By the 1950s, thousands were employed by plants at Oak Ridge and Paducah. The TVA's major users of power, these plants manufactured the bombs that were dropped on Hiroshima and Nagasaki.

Unlike its imitator, the North of Scotland Hydro-Electric Board, the TVA still exists and acts as a wholesale supplier of electric power to over 150 utilities and corporations. Work at Paducah and Oak Ridge continues, the US military having found the Tennessee labour force suitably patriotic and compliant to continue manufacturing nuclear weapons there.

What would Johnston think about such twisted outcomes, and indeed about the ultimate fate of the NOSHEB, swallowed by the Thatcherite privatisation frenzy of the 1980s? Would he have supported New Labour? What would he have had to say about the Scottish Parliament? He was a fervent Scottish patriot, but very much a Unionist. What would he have thought about the Scottish Renewables Obligation which rewards individual landowners by paying favourable tariffs for current generated by hydro-schemes built on their properties. I imagine that he would lament the Obligation which denies Scotland the opportunity for the systematically incorporated development of national hydro power resources. Such futile speculation only reminds us of how the worlds of politics and public service have changed since the days of Tom Johnston. His was a career devoted to the public good and informed by humane concern and egalitarian principle.

At the height of *Forward's* popularity, Johnston was a political idol to much of the British artisan class. In Scotland, at the peak of his popularity, he would be approached on the streets of Edinburgh or Glasgow by people who just wanted to shake him by the hand. He was affectionately dubbed 'The uncrowned king of Scotland'. Hamish Mackinven, who worked for the Board when Johnston took over as Chairman, remembers his own father was one of the many Scots who revered him. Like me, Hamish deplores the fact that Johnston is now largely forgotten, in Scotland and beyond. I am indebted to him for his fascinating account of the great man's working practices.

While mutual respect between the two men made for a cordial professional relationship, it remained rather formal. In spite of his great national popularity, Johnston had a relatively small circle of close friends; he was also a devoted family man. He had a reputation as a bit of a dandy, but his executive style was modest in the extreme. The Chairman may have arrived at the Board's HQ in Edinburgh by chauffeur-driven car, but he would then open his battered old attaché case and set up shop at one end of the board- room table, on the basis that 'part-time chairmen don't need a room of their own'. In fact he did not even take a wage for the work he did at the Board.

In 1945, with his post-war objectives apparently secured, he left politics hoping for a peaceful retirement writing history at his home in Fintry. However, by 1947, he had realised that the NOSHEB needed his continued presence in order to survive. When Johnston did leave the Board for good in 1959, Mackinven was struck by his manner of his departure. 'He didn't say a word to anybody but one day he was driven home to Fintry and he never came back.' There were no celebrations or press coverage of his final retirement from being figurehead of the Board: a quiet exit for the great publicist.

One of Mackinven's anecdotes re-emphasises Johnston's commitment to the working man. By the early 1960s, Willie Logan – Muir of Ord contractor, founder of Loganair and one of the most colourful figures in the whole NOSHEB story – was becoming well-known as a major player on the Scottish construction scene. In 1963, he grabbed the national headlines when he flew into Dundee to submit a tender for the construction of the Tay Bridge – arriving with fifteen minutes to spare, he got the job. Earlier, however, the Highland tycoon had faced a cash flow problem during his firm's completion of the Meig Dam on the Conon scheme and found himself unable to meet his weekly wage bill. Logan had appealed to the Board for help and the Chairman immediately authorised the Board's accountants to provide the hapless contractor with enough cash to deal with the emergency. In making this decision, concern for the welfare of construction workers and their families would have been a factor. Even if Johnston's ultimate goal of full employment in the Highlands was not met, his commitment to the people of the region was unwavering.

I asked Mackinven if he had enjoyed working alongside Johnston. He admitted that Johnston's manner could be dry but he had quite a pawky sense of humour. Despite his family connections to the licensed trade, Johnston was a teetotaller. But the story goes that visiting representatives of the TVA, on a tour of NOSHEB schemes, were taken to a hotel for refreshments. When Johnston asked what everyone would like, the Board men dutifully asked for soft drinks. Imagine their feelings when they heard the Chairman rounding off his order with, 'And I'll have a large dry sherry, please.'

We have seen how much the NOSHEB's creation owed to highly fortuitous political circumstances. Almost as important to the success of the project in its crucial opening phase was the presence on the Board of two hydro boys, expert enthusiasts whose technical skills made major contributions to the work of the Board. James Williamson, who played such a useful part on the Cooper Committee, was one of them and Edward MacColl was the other.

CHAPTER 4

Two Heroes of Hydro Power

James Williamson, CBE, MICE, *1881-1953*

JAMES WILLIAMSON IS ONE of the most impressive and alluring figures in the history of Scottish hydro-electric development. A hydro boy *par excellence*, his was a life of single-minded creativity conducted with great diligence and modesty.

To summarise a person's life in terms of whatever hindsight judges to be his or her supreme achievement can be misleading. We must remember that the whole of Williamson's career as a civil engineer included many more achievements than those connected with hydro-electric development. A glance through the 'Outline of Career' which James Williamson compiled himself in 1951 gives a fair idea of the range of projects which civil engineers routinely design and oversee. His apprenticeship to the Lanark County Council Municipal Engineer was taken up with bridges, roads and the local sewerage system. His indentures completed, he went on to work for private firms. He surveyed and supervised railway development and the refurbishment of Ardrossan Harbour. He designed steel-framed buildings including the Savoy Theatre in Glasgow and laid out a rebuild of the Gleneagles Hotel Golf Course in 1913, using recycled industrial waste from Colville's Dalziel steelworks. He designed power stations, docks and public swimming pools. He also worked on flood prevention measures in the Central Belt and designed and supervised the construction of shipbuilding facilities on the Clyde. Having established that James Williamson's professional experience encompassed far more than just hydro-electric development work, let's look at his biographical background to learn more about the making of a hydro genius.

At first sight, his family circumstances do not seem likely to nurture an engineering maestro. The Williamsons had worked Westfield Farm, Holytown in Lanarkshire since the days of James's great-grandfather, James Williamson, born in 1790. James's father disapproved of his eldest son's break with the family farming tradition and would only finance one year of his engineering studies at university. But Williamson's hydro-electric genius doubtless owed something to his farming background. He remained a countryman all his life, familiar with the natural landscape and very knowledgeable about it. He knew the names of wildflowers and loved walking and climbing. His ability to analyse a landscape and decide on the best way to exploit its hydro-electric potential must have owed more than a little to his farming boyhood and the same could be said for his immense capacity for hard work.

Scotland itself may also have had a critical impact on the imagination of a growing boy at the end of the nineteenth century, when the country was home to a serious concentration of heavy industry. As well as mining for coal and ironstone and state-of-the-art-steel making at Beardmore's and at Colville's, the Central Belt was also the site of some awe-inspiring civil engineering works. The Forth Railway Bridge was the largest in the world when it was completed in 1890. Shipbuilding facilities commissioned in this period would enable Scotland to rank alongside the USA by 1913. Such a concentration of technological advances must have made an impression on the farmer's eldest son.

He did well at primary school and went on to Uddingston Grammar School where languages and maths were his best subjects. His first steps on the hydro trail may have come at Glasgow University where he matriculated in 1897. He acquitted himself well at university, attending classes regularly even in a second year of study at the Royal Technical College (later Strathclyde University) when he had to support himself working as a purser on the Clyde steamers and giving classes at night. He certainly would have been impressed by one of his Glasgow University lecturers, Baron Kelvin of Largs, Professor William Thompson. This great practical scientist was then fresh from his recent triumph, supervising the commissioning of the British Aluminium Company's smelter there at Foyers. The works was powered by Scotland's first large scale hydro-electric plant and every engineering student must have been well aware of the significance of Thompson's achievement at Foyers, especially given the growing importance of electrical power throughout the developed world at this time.

After completing his apprenticeship, Williamson worked for engineering firms in Scotland and England. It is possible to discern in the range of projects on which he was engaged elements that would re-emerge usefully in his hydro power achievements later on, like his wide ranging use of concrete. In 1913, he returned from England to rejoin the Glasgow firm of Formans and McCall, a move prompted by his marriage to local girl, Janet Allison, in the August of that year.

The First World War led to a civil engineering boom in Britain and soon he was engaged on war-related work all over Britain. After a spell surveying at Nobel's explosives factory at Ardeer, he moved to the government munitions factory at Gretna where he supervised the construction of over 80 miles of railway track. He was then seconded to the Royal Engineers and worked on military installations all over southern England. Mrs Williamson, now with a young family, had to accustom herself to the nomadic imperatives of her husband's profession. There would be plenty more yet. After the war, Williamson had a brief spell of private practice on his own in London. However, the venture did not prosper and in 1922, he joined the prestigious firm of Sir Alexander Gibb and Partners as Chief Engineer. It was at Gibb's that James was given the chance to specialise in hydro-electric power generation, a chance he certainly made the most of.

Sir Alexander Gibb (1872-1958) was the product of one of Scotland and Britain's most famous civil engineering dynasties. His great-great-grandfather, William, (1736-91) was a master mason whose son John (1776-1850), worked as Thomas Telford's second-in-command on bridge and harbour construction in Scotland, and was a civil engineering contractor on his own account. John's son Alexander served an apprenticeship with Telford and went on to work with Robert Stevenson. Alexander also continued the family contracting tradition. Alexander's son, Easton Gibb, was Sir Alexander's father and the civil engineering firm he established, Easton Gibb and Co., became one of the most successful in Britain. Alexander joined the family business in 1900, spending 1909-16 on the construction of the Rosyth Naval Dockyard. A succession of wartime government contracts, including extensive work on the French battlefields, earned him the rank of

Brigadier General. His war record brought him greatly enhanced professional prestige and the promise of international contracts at the end of hostilities.

Sir Alexander was particularly interested in hydro-electric power and was aware of its economic potential, not least in his native Scotland. Sadly, Williamson and Gibb never achieved a close working relationship; Williamson's modest self-sufficiency did not suit Gibb's taste for extrovert and distinguished company. Williamson refused to be a social climber. However, it was Sir Alex's keenness to exploit the rising global interest in hydro-electric power which enabled James Williamson to make it his own area of expertise. While at Gibb's he prepared reports on what were then called 'Water Power' schemes in various locations including India, Newfoundland and Norway. Williamson also acted on behalf of the British Government checking the design of hydro-electric schemes in Wales.

In 1925, two dams collapsed in North Wales. They were part of the Dolgarrog scheme owned by the British Aluminium Company. The foundations of the larger, upstream dam had been weakened by unseen water seepage and the damage went unnoticed because the dam had not been kept under professional supervision. The inevitable wash-out struck, with the November rains carving a 50-foot hole in the dam. A resulting mass of water was free to pour down the valley, where the lower dam stood no chance of withstanding the torrent. The accumulated volume of two reservoirs killed sixteen people in Dolgarrog. A dramatic but less serious case of overtopping occurred shortly afterwards when a small storage reservoir above the Firth of Clyde overflowed into the valley below, leaving a trail of destruction. No lives were lost as a result of the Scottish dam failure, but the public concern provoked by both incidents resulted in the Reservoirs (Safety Provisions) Act of 1930.

The Act made Home Office supervision of dam design, building, maintenance and inspection statutory with stage-by-stage approval of dam construction a legal requirement. Since the Act was passed there have been no dam failures in Britain, thanks in part to the government scrutiny of dam building and the control exercised by civil engineers like James Williamson. The importance of such scrutiny was confirmed in the late 1950s by tragically high death tolls following dam failures in France and Spain. Water, a powerful servant, can become a deadly hazard.

Williamson's greatest achievement at Gibb's was the part he played in the promotion and construction of the Galloway hydro-electric scheme in the 1920s and 30s. He was responsible for mapping out the various permutations of design and layout required to secure Parliamentary approval in the face of fierce opposition from amenity and fishing interests. Gibb encouraged him to expand his interest in hydro-electric power generation with an instruction to survey the hydro potential of the whole of the British Isles 'in his spare time'. This is exactly what he did, regularly working away from home and family at weekends. He was also requested by Sir Alex to brush up his languages and to travel in order to keep abreast of the latest developments in the field of hydro-electric power generation. He attended and participated in conferences about hydro development in Europe and America while he was at Gibb's. From 1923 he worked on successive plans for the Galloway scheme and when Parliamentary approval was finally secured in 1929, it was his designs and arrangements for dams, reservoirs and power stations which had won the day. By the winter of 1935-36, the completion of the scheme was finally in sight.

Williamson's success did not, however, earn him the respect of his chief. Gibb still refused to consider him for the partnership that many felt he had earned in full measure. In 1934, Williamson had disagreed with his boss, in front of clients, about the design of Clatteringshaws dam on the Galloway scheme, and although he subsequently apologised by letter, this breach of hierarchical etiquette did not bode well for his prospects at Gibb's. He began to consider taking another shot at private practice; the major project on which this hope was based was the hydro-electric development of Loch Sloy.

Sloy, a loch 800 feet above Loch Lomond with a catchment area 50 miles north of Glasgow, had been the subject of hydro-electric investigation since as early as 1904. Further schemes were mooted for Sloy in the years preceding World War One but none could attract financial support. Later on, inspired by the Snell Report, more reports were made on Sloy. But local opposition combined with continuing financial uncertainty to prevent anything being done.

By 1937, with his hopes for advancement at Gibb's more or less defunct, the development of Loch Sloy by the promoters of the Galloway scheme, the Power and Traction Finance Company, offered Williamson a sound basis on which to set up on his own. Based in Westminster, Williamson compiled all the material required for the promotion of the scheme in Parliament: maps of the area, plans and sections of the intended scheme and lists of the relevant landowners.

However, pressure from the coal industry and many MPs' desires to safeguard the Scottish landscape caused the promotion to be defeated. Hydro power was rejected in favour of a coal-powered station at Yoker on the Clyde. Had the promotion succeeded, Williamson's solo career would have been safely assured. But although he was able to pick up some work, in 1938 he accepted an invitation to join Sir William Arroll and Company as Director and Chief Engineer with express permission to undertake private consulting work in his spare time.

At Arroll's, he took on the usual range of projects, many inspired by the Government's re-armament plans. But the recognition and flexibility granted to him at Arroll's meant that in 1941 he could take his place as Technical Adviser to the Cooper Committee. Here he was able to state case for hydro power development and at last a supportive House of Commons accepted the arguments. The Committee's eventual achievement, the passing of the Hydro-Electric Development (Scotland) Act in 1943, meant that James could be sure of regular employment because of his success at Galloway and the work he had already done on Sloy. Most important of all for his future prospects was his position as Technical Adviser to the newly-created North of Scotland Hydro- Electric Board. This appointment would surely put him in a good position to secure plenty of work designing and supervising the Board's Development Plan. Indeed, although the Board instituted regulations to ensure that Board members did not use their positions unfairly to gain work for their own firms, James Williamson was never without significant and rewarding employment until his death in 1953.

In 1944, with the promise of employment as Resident Engineer at Sloy, he left Arroll's and started a private practice in Glasgow. (A Resident Engineer is employed by the consulting engineers who design hydro schemes to oversee the contractors on site.) In 1947, the firm of James Williamson and Partners was set up. In 1989, the Williamson Partnership, making a realistic assessment of the firm's position in the contemporary civil engineering scene, undertook a successful merger with Glasgow's Mott MacDonald Ltd. The partnership's area of expertise within the new group was power generation and John Cowie, Senior Partner of the new firm wrote in 1989 that James Williamson would have certainly approved this destiny of the firm he had founded nearly half a century earlier.

Williamson's wife and four daughters, as the mores of the first half of the last century dictated, seemed not to have objected to his passionate involvement in his work. He would often be away working at weekends especially after Sir Alex's order to survey the whole of Britain's hydro-electric potential. Molly, Williamson's second daughter, reminisced in 1957 for Williamson and Partners in-house magazine about the years her father spent preparing and executing the Galloway scheme. 'Almost all his holidays', she recalled, 'were busmen's holidays'. She remembered with surprising pleasure a whole month that the family spent in Galloway during one of the wettest summers on record: 'We sloshed about day after day, inspecting rain-gauges on rain-swept moors, while the midges slowly devoured us'.

Williamson recruited Stan Young in 1943 because of his skills as a Home Guard map-maker. He needed Young's help with the documentation of the NOSHEB Development Plan. Stan remembered an occasion when Williamson had surveying work to do in the Findhorn-Spey area and announced to his staff that he thought 'the family wish to holiday at Nethybridge this year'. During the ensuing vacation, Williamson would be dropped on site every morning by his family who duly returned to fetch him in the evening.

More dramatic was Molly's story of Williamson's near arrest in war-time after his activities around Pitlochry had attracted police attention. His maps covered with 'cryptic markings' confirmed their worst suspicions and he came very close to being arrested as a spy.

His family adored him despite his concentration on his work and his staff did too. Indeed he treated them like a second family, buying the 'girls in the office' glazed fruit at Christmas and addressing even the lowliest employee by their full title: Mr Young or Miss Browning. He died in 1953 from a smoking-related illness;

there are plenty of affectionate references to the trails of cigarette ash left as the boss moved about the office and the private moments he enjoyed with a 'proper cigar' in his own office. There is also the possibility that the huge amount of physically and mentally demanding work he accomplished in his career of over 50 years might have contributed to his relatively early death.

Few photos of him survive but the picture of him cutting the breakthrough tape in the main tunnel at Sloy provides the perfect image of a modest man overjoyed at the realisation of many years spent working for the implementation of a comprehensive hydro-electric power generation scheme in Scotland.

James Black was a 23-year-old General Foreman at Sloy. The GF was in overall charge of the men on site on behalf of the contractors. James Black was given the job of escorting Williamson round the site when he came to join in the celebrations for the tunnel breakthrough. A breakthrough happens when the two teams of tunnellers who have been working from opposite ends of the tunnel meet in the middle. Black remembers Williamson as a frail old man whose wide-brimmed hat had seen a lifetime of 'sloshing about' in the rain, and who had nothing but praise for the labourers who had executed his plans.

Sloy was designed by Williamson to be the uniquely important flagship of the whole NOSHEB project. He had been aware of Sloy's hydro-electric potential for years and he was not alone; Edward MacColl, NOSHEB Deputy Chairman and Chief Executive, had already tried to develop Sloy when he was employed by the Central Electricity Board in the 1930s. Like Williamson, MacColl's presence on the Board was highly fortuitous. The NOSHEB could certainly never have been born without Tom Johnston but the quality of its founding achievements owed a great deal to the work done in the Board's opening years by Williamson and MacColl. Let's now have a closer look at the career of Edward 'Electricity' MacColl.

(This section on James Williamson could not have been completed without reference to two private publications: *Williamsons* by Robert Buchanan and *Venture and Win* by Donald Hamilton. I am very grateful to Tom Douglas of Mott Macdonald Ltd Glasgow, for making them available to me.)

Sir Edward MacColl, 1882-1951

The engineering career of Edward MacColl, like that of James Williamson, did not start with the benefit of parental support, MacColl's stepfather having declined to fund his higher education. In 1901, after completing his engineering apprenticeship with John Brown and Company, the Clydebank shipbuilders, MacColl went to work for the Glasgow Corporation Tramways Department. The corporation's system underwent electrification in the same year and MacColl started work for the Department at Pinkstone Power Station. He soon gained a considerable reputation for his work and this led to his being recruited by the Clyde Valley Electric Power Company in 1918 as Chief Technical Engineer.

At this point, the CVEPC was the biggest generator and distributor of electrical power in Scotland, with thermal stations at Yoker, Clyde's Mill and Motherwell. This pre-eminence made the Company the obvious choice to undertake the hydro-electric development of the Falls of Clyde. The Power and Traction Finance Company, financiers from London, were the promoters of the development. That they were successful in obtaining Parliamentary approval for their scheme owed a great deal to the meticulous attention which MacColl had paid to its preparation. He made strenuous efforts to deal productively with all the riparian owners affected by the scheme and he was deeply attentive to the amenity issues involved in the development of one of central Scotland's favourite beauty spots. He decided that a run-of-river scheme was feasible, thereby obviating the need for dam or reservoir. He ensured that the scheme's two power stations, at Stonebyres and Bonnington, were designed to blend harmoniously with the surrounding countryside. He also made certain that ample compensation water would be available to preserve the attractions of the famous Falls. The scheme, Scotland's first extensive hydro-electric public supply scheme, worked effectively in conjunction with the company's existing thermal stations. Despite the derision of the coal lobby, it was a great success and served as a pioneering example of how well hydro-electricity could be harnessed for public power supply.

MacColl's next appointment, in 1927, was to the Central

Electricity Board, recently instituted as a result of the Weir Committee's recommendations. MacColl was appointed CEB Engineer for Central Scotland. His first project there provided the Scottish hydro-electric industry with the practical means it needed to achieve economic justification. This crucial achievement was the long-distance transmission of electric current. MacColl himself laid down the technical foundations of this development with the invention of the MacColl Protective System that would be used all over the world for the control of long distance transmission lines. By 1933, MacColl had supervised the creation of the Scottish National Grid, Britain's first regional electricity supply Grid. For the required interconnections to work, MacColl had standardised the frequency at which local current suppliers operated to 50 cycles per second. With the lowest administration and maintenance costs in Britain, the Central Scotland Grid was to serve as a model during the creation of the rest of the National Grid. MacColl worked for the CEB throughout the 1930s. In 1936 he submitted a proposal for a pumped storage hydro-electric scheme at Loch Sloy, but the CEB turned it down. On the outbreak of the Second World War, MacColl was put in charge of the spares pool for electric power generation and distribution. Increasingly, he felt that his work at the CEB was not calling for the sort of innovative creativity at which he excelled.

In 1943, however, these professional frustrations vanished with an invitation to join the North of Scotland Hydro-Electric Board as Vice-Chairman and Chief Executive Officer. It was an inspired choice; as well as his great technical expertise and inventiveness, MacColl had a proven ability to deal decisively with any opposition that he encountered. 'Let's give him the dull thud' was his frequent response to anyone obstructing his plans. His other personal catchphrase, 'Let's get on!', signalled his boundless enthusiasm and determination to proceed with the project in hand. He particularly loathed any interference from London in the Board's plans. He was a small man and was often heard to complain about the preponderance of tall people in the English capital. Like Johnston, he wanted to protect the Board's operational independence at all costs. His expert abilities and determined commitment were precisely what the NOSHEB needed.

MacColl agreed with Johnston that hydro-electric development offered vast potential for bringing prosperity to the Highlands. As a devoted patriot, MacColl felt keenly the importance of rescuing the Highlands from isolation and poverty. He had a strong sense of the destabilising threat that economic depression posed to civil society and knew that the decline in Scotland's manufacturing base since the Depression of the 1930s could only exacerbate dangers such as fascism. MacColl saw hydro-electric development as a force for social good.

The NOSHEB development at Sloy was forced to submit to a Parliamentary Inquiry and was only carried through Parliament after certain modifications were made to the Board's original plans. However, despite the vehement opposition faced by the Board over its intentions at Pitlochry, MacColl insisted that work there should not be postponed. The scheme, the enlargement of the existing Grampian Electric Supply Company works on the River Tummel, was essential to the overall economic strategy of the Board and its plans to provide cheap connections throughout the Highlands. In the event, the Pitlochry scheme was authorised by Parliament without any changes having to be made by the Board. The threats and complaints of the Pitlochry opposition evaporated. Thanks to MacColl's characteristic determination, the NOSHEB pushed on successfully with their Pitlochry development and their overall plans lost no impetus at all.

One of MacColl's most outstanding contributions to the Board's execution of its Development Plan concerned the appearance of the power stations built as part of that Plan. An architect, James Shearer, employed by the Board to design a couple of small power stations, at Morar and at Kyle of Lochalsh, was impressed by the virtues of what he called the 'cottage architecture' of the Highlands and wished to use locally quarried stone. Unsure of the Board's reaction to the extra time and money his idea might require, Shearer rather diffidently approached MacColl for his approval. In fact, MacColl, whose constructive imagination ranged very widely indeed, had already been considering this very option for power station construction. Although most engineers would only see a power station as 'a box to hold machinery in', MacColl had already realised the political importance of a power station's looks. A handsome power station, designed to blend with the natural landscape, would be valuable in fending off accusations of environmental vandalism from the amenity lobby.

To show Shearer why he, MacColl, thought local stone was potentially very useful for the NOSHEB's purposes, he took the architect and other members of the Board to see a dam constructed using concrete some years before. It demonstrated perfectly the flawed appearance that concrete takes on after decades of exposure to the elements. MacColl did not simply rubber-stamp Shearer's ideas about local stone and leave it at that. Instead, he asked the architect to establish where the NOSHEB contractors would be able to procure supplies of native Highland stone. MacColl then made a point of placing substantial orders with some of the quarries identified by Shearer, to make sure they and their associated craftsmen survived in business to play their part in the NOSHEB Development Plan. Many NOSHEB power stations were built either entirely of sandstone or of ferro-concrete (concrete reinforced with steel) clad with the same stone.

The Board's use of local stone was a magnificent idea and produced many fine looking and harmonious buildings, including some handsome accommodation for the Board's permanent staff throughout the Highlands. The sandstone has weathered most attractively. Much of the stone used by the NOSHEB contractors came from Greenbrae Quarry at Hopeman near Elgin and Tarradale Quarry, near Muir of Ord, which was owned by Willie Logan's family at the time of the Development Plan. The local stone-quarrying and stone-cutting industry received a great boost from the Board's activities thus achieving one of MacColl's objectives for local enterprise.

Fasnakyle Power Station, designed by Shearer, is now a listed building. It was opened in October 1952 by the Duke of Edinburgh. The Duke confessed at the time to having originally been concerned that the worst accusations levelled at the Board by the amenity lobby might be true. But he was delighted with Fasnakyle. Shearer had created a frieze of circular relief sculptures to line the top of the power station walls inspired by Pictish and other ancient carvings he had seen in the Highlands. (By the 1960s, power station design reverted to strictly utilitarian principles when the technology became available for turbines to be operated in underground structures.)

MacColl was also interested in folklore and mythology. In a memorial volume of essays commemorating her husband's life and career, his widow, Lady Margaret, recalled his keen independence of mind. To illustrate the wide range of her husband's interests, she remembered that, having repeatedly refused invitations to become a Freemason, he eventually joined the Viking Society and the Newcomen Club as well as being an active member of his clan. And in addition to being fascinated by the past, he was keen to experiment with technologies of the future. While he was busy supervising the myriad details of the NOSHEB Development Plan, he was also always looking for new ways to manufacture energy in the Highlands. He initiated research into the possibility of wind power with a windmill on Orkney and he had hopes of a peat-fired turbine at Altnabreac in Caithness. Neither of these experiments succeeded, but MacColl had established the importance of non-hydro power generation which was to play a significant part in the Board's future activities.

One of the high points in his distinguished career must have been the ceremonial opening of Sloy. He took characteristically detailed care of the arrangements. He erected covered stands and put a light bulb under each chair to keep the spectators warm. James Black, who was present, recounts that these stands blew over in gales the night before the opening, but MacColl was able to make sure all was in order in time for the big day. The successful completion of the construction at Sloy and the inauguration of its commercial operations were as much of a dream come true for MacColl as for James Williamson.

'Electricity MacColl' was knighted in 1949. His death two years later deprived him of the chance to enjoy the honour and his planned retirement projects in full. He had intended 'working on the land' and adding to his knowledge of classical music after he left the Board. Many of those who knew him, including his widow, were sure that his death had been hastened by overwork. During his nine years as overall boss of the NOSHEB, his actions were ruled by two key objectives: to safeguard the independence of the Board and to maintain the impetus of its plans for the regeneration of the Highlands. After his death, Lady Margaret wondered to what extent his last years had been clouded by severe anxiety over the very survival of the NOSHEB and its founding achievements. His wife was accustomed to the intensity of his engagement with his work; MacColl regularly covered tablecloths and walls with his calculations. But she also remembered him saying, towards the end of his life, 'Hydro-electricity is not enough'. Was it possible that, even in 1951, MacColl realised that the problems of the Highland economy would ultimately be too intractable for the provision of electric power to solve?

There can be no questions, however, about MacColl's decisive contribution to the work of the Board. His determination to overcome any obstacle and his innovative imagination were invaluable. Sadly, he did not live to be present at the official opening of the Board's new works at Pitlochry or to see the almost instant success of the visitor centre adjoining the dam, where salmon can be observed through a viewing chamber as they negotiate the fish ladder on their journey upstream. It was Sir Edward who decided that the dam at Pitlochry should feature a public walk and that a visitor centre should be opened there. He also insisted that Pitlochry Burgh Council should be allowed to choose the name of the reservoir created by the new dam. Faskally was the Council's choice, being the name of a local property opposite Clunie Power Station. The fears of the local hoteliers about the dire effects that hydro development would have on their trade have been totally disproved; last year 120,000 people visited the centre. I hope they took note of the memorial plaque:

SIR EDWARD MacCOLL 1882-1951
ENGINEER AND PIONEER
HYDRO ELECTRICITY SCOTLAND

Like Tom Johnston, MacColl played a vital part in bringing the Highlands into the twentieth century and apart from this plaque and a street name or two, his contribution has been virtually forgotten in the North.

CHAPTER 5

Glorious Years: The NoSHEB Development Plan in Action

Sloy 1945-50

SLOY'S STRATEGIC ROLE IN the whole NOSHEB project gives it a key place in the story of hydro-electricity in the Highlands. Williamson and MacColl had spent thousands of hours investigating the scheme which formed a vital part in the launching and survival of the North of Scotland Hydro-Electric Board.

I was lucky enough to meet James Black who worked as a General Foreman on the Sloy scheme. He began his construction career in his native County Antrim but came over to Scotland to search for more work and higher wages. He went on to a highly successful career in the industry which included work for NOSHEB contractors at Fannich, Invermoriston and Lairg and the supervision of several hydro installations.

Peter Payne's study, *The Hydro*, is a wonderfully informative book which cannot be faulted for its technical descriptions and economic analyses. However, listening to James Black, confirmed for me the value of hearing about things straight from the horse's mouth; he gave me insights into Sloy and the later NOSHEB schemes which no amount of reading could have done. He is the only person I have been able to talk to who worked at Sloy for any length of time. Not only does he have a knowledge of the construction world born of a lifetime's experience, he also speaks with the special insight of someone who could say that Sloy had been 'the best years of my life'.

Remembering how close the British people had come to losing everything to Nazi Germany only five years before work started at

Sloy helps us to realise what a special atmosphere existed throughout the whole country as the huge and exciting work of national reconstruction commenced in 1944. This was not just a time of rations and shortages and plenty of folk still away from their families. It was also a time of inspired determination among all sections of the population to work together to re-build society. What better and more visible way was there to achieve such aims in the north than by the construction of major civil engineering projects which would employ thousands and improve the standard of living for the Highland population?

The task of exploiting the hydro potential of Loch Sloy and its catchment area was a huge one though the amount of constructional detail given here does not indicate any singularity about the way the scheme was designed or built. All hydro-electric installations represent the civil engineer's individual solution to a set of challenges posed by a particular terrain and the precise generating requirements forecast by the current supplier. All schemes are therefore unique and noteworthy. But looking closely at Sloy does enable us, with our narrative poised at the start of a decade and a half of NOSHEB construction work, to consider in sequence all the stages involved in the construction and commissioning of a hydro-electric power generating scheme.

Moreover, Sloy, as the Board's first project, had an economic and political importance which cannot be overstated. Loch Sloy was the obvious choice for the Board's first major undertaking. It promised to satisfy the Board's urgent revenue requirements, thanks to a relatively straightforward connection route to Glasgow and that city's growing demand for electric current. Johnston had committed the Board to supplying electricity to the Highland population as a public service rather than a profit-making venture. Such a socially-justified policy was to be financed initially by the sale to Glaswegian consumers of current produced at Sloy.

To this end, a dam would be built to store water in Loch Sloy and this water would be piped through the middle of Ben Vorlich to a power station on the west shore of Loch Lomond. The current generated there would be transmitted to Glasgow via high voltage cables strung on pylons.

How simple this plan looks written down, but at that time and in that place almost nothing turned out to be simple. The Board, eager to assert its national usefulness by making a significant contribution to the post-war fuel crisis, announced that Sloy would be producing electric power by 1947. In fact, the scale of problems encountered by the contractors meant that Sloy was not completed until 1950.

The end of the war left Britain short of almost everything; raw materials and labour for reconstruction were in great demand throughout the country. There was still relatively little mechanisation of the tasks involved in hydro power development. These tasks were common to all the Board's construction schemes and generally occurred in the same sequence, one that we will now examine.

The decision to develop a hydro-electric generating scheme in a particular place can only occur after civil engineers have made extensive surveys of the area. As well as recording precisely the shape and geology of the terrain, they must produce estimates of rainfall and run-off. Run-off is calculated as the amount of water leaving an area via its rivers and streams. These findings, together with the analysis of all relevant economic factors, dictate the options open to the designers of hydro-electric power generating schemes.

Once the site at Sloy had been selected and surveyed, access to it had to be established for the delivery of materials, machinery and other equipment. The extreme difficulty of gaining access at Sloy was not unusual; NOSHEB schemes are typically located in wild and relatively inaccessible places. Once access had been provided, accommodation for the labour force could be set up and a temporary power supply from a diesel generator installed. The contractors assembled earth-moving plant and set up a maintenance workshop to service it. A weigh-batching plant was also built which measured and mixed the materials used for making concrete: crushed stone, cement and sand. These ingredients were delivered via the new access roads.

With these basic facilities in place, construction could start. Huge earth moving operations preceded the laying of the dam's foundations. Because of the tremendous threat to human life and property posed by dam failure it is crucially important to ensure that dams have secure and durable foundations. With the foundations complete, the joiners and concreting crews could begin the construction of the rest of the dam; many NoSHEB dams, like Mullardoch which towers 150 feet above its foundations, are huge structures. Supplying materials to dam sites and organising the labour force on them are similarly huge operations.

Tunnelling was an essential element of all the NOSHEB schemes; 200 miles of tunnel would be driven during the implementation of the Development Plan. Tunnels are used to take water from one valley to another if generation operations are easier to carry out there. Tunnels are also used to collect water from side streams. Such straightforward transfer requires aqueducts that divert and carry freely flowing water. But for the delivery of water to the turbines, high pressure tunnels or penstocks are needed. They take water, under pressure, directly to the power station. These tunnels have to have special design features to withstand the huge stresses at work when water flows intermittently under pressure. Engineers counter the destructive effects of pressure-related stresses by looking at a tunnel's cross-sectional shape and the way it is lined. However, such specialist attention to design and construction can increase costs significantly.

All tunnelling is potentially costly, difficult and dangerous. Whatever the complexities of their design, all the NOSHEB tunnels posed the problem of blasting and mucking. Hundreds of thousands of tons of rock were broken by the blasting of the tunnellers. The resulting spoil had then to be removed to the surface. Tunnels are perilous places, however well organised the workforce. Handling gelignite is a hazardous business and its detonation produces carbon monoxide in significant quantities. Further lethal dangers underground include the ever-present possibilities of random rockfall and water seepage. We will learn more about the workers' experience of these hostile conditions later.

By the time that construction work at Sloy was finished, the mechanisation of many of these essential hydro jobs had advanced significantly from the picks, shovels and wheelbarrows used at Foyers and Kinlochleven. Official tunnelling rates in Lochaber in the 1920s had reached 90 feet per week. By 1945, mechanical excavators and lorries were saving contractors thousands of man-hours and tunnelling rates at the end of the NOSHEB construction programme had increased nearly six-fold from those on the first

Highland hydro tunnels. By 1948, a quieter and quicker drill had been introduced for making the openings in the rock where the gelignite charges were inserted. It had a tungsten carbide tip and replaced the pneumatic drifters which had been in use since the mid-nineteenth century The new drills made for dramatic increases in work rates. A drifter had taken three men half an hour just to set up before drilling could begin. The drifters' mild steel bits would be blunted and changed several times before one hole was finished. Some of the new drills could be left to work, two or three at a time, while their operator set up more to eat into the tunnel face.

Not all the contractors employed by the Board on its Development Plan supplied their men with the most modern machinery available. Drifters were used by the men who drove the tunnels from Mullardoch to Fasnakyle on the Affric-Beauly Scheme. But even when the newest machinery was used, hydro construction was still massively labour intensive.

Progress at Sloy was hampered by a dearth of specialist workers, such as carpenters. The carpenters' essential role was to make the shuttering for concrete dam construction. (Shuttering is a temporary structure, usually timber, which contains and shapes concrete while it sets.) Tunnellers, also known as rock miners, were particularly hard to find. Coal miners refused to bring their underground skills to the driving of the Board's tunnels. The coal miners considered hydro tunnels unsafe, as they were not reinforced by the timber supports compulsory in coalmines. To solve this shortage of specialist skills at Sloy, training schools were set up there giving instruction in tunnelling and carpentry.

Despite these training schemes, finding sufficient manpower for the job of building the scheme at Sloy continued to prove very difficult. The Board made desperate requests for workers to the Ministry of Labour but demobilisation was far from complete in 1945 and there was an intense need for construction workers everywhere in war-damaged Britain. One solution was the employment of German Prisoners of War. The British Government sanctioned the employment of German POWs on various domestic reconstruction projects for some time after Germany's surrender. The British public did not seriously question this at first, though implications for the Geneva Convention were somewhat unclear.

By 1947, mounting public protest forced the Prime Minister, Clement Attlee, to speed up the repatriation of German prisoners.

There was, however, another group who would make a crucial contribution to the Board's work at Sloy and to the rest of the Development Plan. Displaced Persons, or DPs as they were called, were men and women who after the political cataclysm of World War Two preferred to make their homes in Britain rather than return to their native countries, many of which were now under Soviet control. Among their number were Estonians, Latvians, Ukrainians, Germans, Yugoslavs, Hungarians, Czechs and Poles.

The DPs were taken on by the contractors at Sloy to work for half pay. They had to prove their commitment to their new country by staying at Sloy until the scheme was finished. The DPs' motivation was, therefore, very strong and as they made up two thirds of the workforce at Sloy their presence, as we shall see, made a huge difference to the way the scheme was carried out. It is hard to imagine where else the contractors could have found the workers they needed at this time.

Materials in any amount and quality were as difficult as labour for the Board's contractors to procure. A bag of cement was 'worth its weight in gold', as one Board employee remembers. Williamson's revolutionary buttress design for the dam at Sloy minimised the amount of concrete needed by simply omitting any material which did not directly contribute to the dam's stability. The design called for 50,000 cubic feet less concrete than a conventional gravity dam would have needed, but even that amount was difficult to secure. Local quarries were accessed for stocks of sand and aggregate. However, local suppliers could not guarantee regular bulk deliveries in the quantity required so the contractors had to take what they could get in bags which meant extra handling and storage costs.

Timber for shuttering was also very scarce. Steel shuttering was tried which worked well for the construction of the dam buttresses. However, steel for lining the tunnels proved hard to find. These materials shortages also intensified labour problems as they delayed the provision of accommodation.

Tom Johnston personally addressed the accommodation problem in a typically single-minded way. Frustrated by the sight of abandoned army camps all over Scotland, he had attempted to arrange for their official acquisition by the NoSHEB contractors. Finding state bureaucracy unresponsive to this sensible idea, he 'decided that the risk of prosecution [for larceny] was worthwhile' and simply sent the contractors to remove the camps *en bloc* for use at Sloy.

The building of Sloy had to overcome more obstacles than those connected with difficult post-war economic circumstances. The physical geography of the crag-bound site presented major operational problems. The rough and awkward terrain around Loch Sloy, Loch Long and Loch Lomond meant that gaining access to the site had been a major undertaking in itself. As well as hundreds of thousands of tons of building supplies, the contractors also had to transport very heavy and extremely delicate electrical components such as the spiral casings for the turbines, weighing 4.2 tons apiece. Special lorries were adapted for this sort of transport problem. They travelled at a snail's pace and were in constant danger of blocking the road round Loch Lomond, at that time the main route to Inverness and the north.

Progress at Sloy was seriously hampered by the constantly vile weather. Incessant rain, driven horizontal by gale force winds, was an almost permanent feature on site. Rain caused special problems in the concreting process and during the three years the dam took to complete, only 21 days without rain were recorded.

The contractors on the NoSHEB schemes paid bonuses to encourage a consistently high work rate. I asked James Black why there was such pressure to complete contracts; as long as the work was finished on time, then surely 'more haste less speed' made sense? He explained by telling me exactly what had been at stake for the contractors throughout the rest of the Development Plan.

'Those contractors had a good reason to hurry. Every day spent working on a scheme cost money in overheads; clerical, catering and cleaning staff all had to be paid while a scheme was under construction.' Just as important was the contractors' perception that a speedy work rate would win them more of the mountain of contracts promised to the industry by the Board's Development Plan. To advertise their firms' prowess, contractors published their tunelling rates in trade journals and national newspapers. Individual firms competed for tunnelling world records.

Black also told me about a practical factor that made Sloy different from subsequent NOSHEB schemes: the presence of the DPs with their uniquely high motivation levels. Unlike some of the Irish who even at Sloy, Black remembers, tended 'to up sticks and away', the DPs had to stay on site if they were to have the chance of British citizenship. Many DPs feared that returning to homelands now under Soviet control would have made them fatally marked men and women. Modris Ekstein's powerful study of the DPs' experience, *Walking Since Daybreak*, contains horrifying evidence of this compelling fear. Russian soldiers who had fought against the Soviet state smashed the windows of the trains about to take them back to Russia so that they could use the splintered glass to cut their throats. DPs at Sloy were determined to avoid repatriation. 'They had to prove themselves fit citizens,' said James, 'and that meant doing their jobs well'. The DPs excelled in the Training Schools and went on to work on the construction of Sloy in a well-organised fashion, which was not only highly productive for the contractors but also very safe for the workers. The tunnelling sequence is perfectly safe if it is done systematically and all the rules, especially those to do with gelignite, are followed strictly at all times. 'The DPs still had war-time discipline', Black told me, 'and they worked together to make sure Sloy was a success'. Only a very few ever disappeared or were returned to the detention centres. The DPs were the 'back-bone of the workforce at Sloy'. With their devotion to the task in hand and their tendency to look after themselves and their fellow workers like the members of an army platoon, it is easy to see why their contribution at Sloy was so considerable.

There were twenty nationalities at Sloy. James told me about playing in the camp football team: 'You know, there were nine different nationalities in that team, including two Yugoslav internationals. But there were no conflicts between groups or individuals, even when they'd been on different sides in the war'. Seemingly incongruous friendships were made between these different nationalities. One that Black recalls vividly was between a big White Russian called Joe from the Ukraine and a little Hungarian. 'They were working together carrying timber for the carpenters when Joe slipped, fell heavily on a stone and lost his pinkie. I took him to the First Aid Tent in my jeep. They staunched the blood and sent the big fella off to hospital. The wee Hungarian came running up an hour later yelling at the top of his voice that he'd got Joe's finger, Joe's finger, Joe's finger... I reckoned that would have been about all the English he had. But he just wanted to help his mate.'

I questioned Black further about this odd absence of national divisions. 'If there was ever any trouble among the DPs, they sorted it out for themselves', he replied and told me a story to illustrate his point. A number of the DPs had been in Nazi concentration camps. They could be recognised by the numbers tattooed just below their hairline. Black told me that this group of DPs didn't say much about what had happened to them in the war but they tended to be shy and withdrawn, especially when they first arrived. Girls from the concentration camps were put to work in the canteens. One of these, Anna-Laura, was dishing out the soup one evening when a German, dissatisfied with what he'd been given, threw his bowlful into the girl's face. 'Four fellas got up,' Black went on, 'an Irishman, a Scotsman and two DPs. They got a hold of this guy, put a rope round his wrists and strung him up from the crossbeams of the canteen. Nobody moved, nobody said a word. He wasn't any trouble after that'. I asked if there had been any awareness at Sloy about exactly what the tattooed DPs had been through in the Nazi camps. 'We'd no idea,' he replied. But the vulnerability of that girl at least had been noticed and, as this vignette shows, acted on.

By 1951, Sloy was in full operation and the NOSHEB Development Plan had been put into action all over the north. Schemes still had to gain Parliamentary sanction and the traditional opposition to hydro development had not simply disappeared. But the 1943 Act, and a general post-war readiness to accept change which promised benefits for the whole of society, meant that the Board was able to implement its Development Plan virtually intact throughout the 1950s. This decade was indeed the heyday of hydro development in the Highlands and saw constructional schemes underway all over the north.

Work

The Engineer was on the beer,
The Shift Boss had the flu.
There was no one left to drill the face
But Mackenzie, Logue and MacHugh.
The EIMCO roared and the shift boss snored
And the gellie reek was cat*.
We finished her round
And we mucked her sound
And that, my boys, was that.

* 'Cat' is a colloquial abbreviation of 'catastrophic' used all over the NoSHEB construction sites.

Some Ross-shire men I spoke to took part in a well-publicised world record attempt on a Hydro Board tunnel at St Fillans in the Perthshire hills. The tunnellers were poised to start drilling through a stretch of 'red lava rock so soft you could just eat your way through it'. So the contractor decided to make an official attempt on the world record. 'We went for it on the Monday and broke it on the Wednesday', one man remembers. The boys drilling at the face were the stars of the show. They didn't even have to muck the sides of the tunnel, as they normally would have done. Men came behind them and took the spoil away in wheelbarrows. To celebrate the men's success, the contractor gave them all gold cigarette cases, lighters and wrist watches, engraved with each man's name and the details of his world record triumph: 'Aye, we were sound!'.

Competition for contracts meant that firms like Cochrane's, Bisset's, Logan's and Whatling's were forced to pay bonuses to encourage progress rates. As we shall see, in the absence of health and safety regulations (serious legislation was not introduced until 1974), this institutionalised haste led to high accident rates throughout the schemes as short-cuts may have been taken, regardless of safety. Men who had been employed on the schemes in the 1950s and 60s told me that their work took place in an atmosphere of danger, especially in the tunnels. At the time, the workers only complained about the riskiness of what they were doing with humorous understatement, for example at the start of a shift in the tunnel: 'It's a bit rough in here the day'. Everyone was too well aware of the financial rewards to let safety worries loom large. 'It was the bonus that made it: you just had to keep going.'

The highest wages were paid to the men in the tunnels and consequently the tunnellers had enormous prestige. They even had a special nickname: the Tunnel Tigers. I asked one veteran Tunnel Tiger if he hadn't been scared when he went into his first tunnel. 'You just went in with the rest. If you were scared you didn't want to show it.' The contractors' men became hydro adventurers, taking maximum advantage of the employment offered by the huge NoSHEB project, which at its height employed over 12,000 men. They were able to move from scheme to scheme, more or less at whim, secure in the knowledge that there was more work down the road or in the next glen: 'When you were browned off with one scheme you just moved to another one'.

Peter Payne's history of the Board describes the spirit of adventure that permeated the activities of the Board, from the chairman down to the 'mud-caked labourer'. Not all the labourers may have been fully aware of Johnston's aspirations for the Highlands, but without doubt Highlanders had never before been offered the opportunity to make so much money right on their own doorstep. Employment opportunities available to most Highlanders in the post-war period were confined to low-paid agricultural or domestic work, with emigration, temporary or permanent, as the only alternative. Like the geniuses of hydro-electric design who looked to make electrical power from the Highland climate and terrain, the labourers on the schemes were opportunists. They braved extremely perilous working conditions to earn wages at levels previously unknown in the Highlands.

Men came from all over the north to work on the NoSHEB Development Plan. There were Highlanders from the east and west coasts as well as men from the Hebrides, often Gaelic speakers. The fishing industry was in serious decline after the war and many men from the fishing villages of Banffshire and Aberdeenshire came to the schemes where they were known as the Seagulls. Scots came from all over the central belt and a few English and Welsh also joined the workforce.

The DPs were never again as numerous on any single scheme as they had been at Sloy, but as individuals their position was now much stronger. They had graduated from Sloy as full British citizens and half pay was a thing of the past. Many had done so well learning the job at Sloy that they could now be taken on as general foremen and tunnel bosses. All of Lairg knew the Germans were 'the kingpins in the tunnel' which the contractors, Wimpey, drove from Inveran to Loch Shin. In these exalted positions, the DPs were able to capitalise on the system of bonus payments, receiving the lion's share of the extra money. The tunnel boss, the shift boss and the leading miner took the biggest share of the wages and so had reason to push the work on. This division of the spoils was the contractor's way of ensuring that the labour force would be compelled by their immediate bosses to respond to the challenges of the bonus system and so maintain their work rate.

Many of the DPs who remained in hydro work were as single-minded as they had been at Sloy, saving their considerable earnings rather than using them up on drink and gambling. Poles and Czechs became well known for their tunnelling prowess. Many former DPs were also known to be very rich men.

The Tunnel Tigers, whatever their nationality, had tremendous status on Highland streets when the Development Plan was underway. People recognised them by the whiteness of their skin and were duly impressed: 'Look, that's one of that Tunnel Tigers'. This reputation rested on the huge wages that these men were being paid as well as the dangerous conditions in which they worked.

The Tunnel Tigers, however, did share one invisible, unglamorous characteristic. They were inclined to suffer from terrible ulcers, a direct result of the hectic shift patterns dictated by the bonus system. After as long as twelve hours without even a cup of tea, the men were often too tired to eat properly at the end of a shift. But they probably found time for a beer or two and so woke in the morning suffering painfully from the effects of the day before. When it was time for work, 'You'd grab a handful of Rennies and away you'd go'. Many of the tunnellers came from Ireland and indeed the Irish made up a large proportion of the whole workforce. The conditions of extreme rural deprivation which had forced generations of Irish people to seek work in Scotland since the beginning of the Industrial Revolution still obtained after World War Two; the descendants of Patrick

MacGill's compatriots at Kinlochleven made a major contribution to the completion of the NOSHEB Development Plan. Many of them came from Conemara and Donegal and their poverty-stricken appearance background shocked some Highlanders, especially those who had never seen toe-rags before. Instead of wearing socks, the Bog-Arabs (as the Irish were called on the schemes) wound strips of old blanket round their feet, only replacing them with a new set when the old ones had worn out. This practice was, for men in wellie-boots working in wet tunnels, more sensible than wearing socks. Toe-rags could be rinsed out at and dried more easily than socks and some Highlanders adopted their use.

Many of the Irish worked seasonally for the NOSHEB contractors, returning home after a winter in the Highlands to work on their smallholdings. All the men tended to stick together according to their geographical origins. Even amongst the northern Scots, different groups like the islanders and the Seagulls kept to themselves. But despite the men's tendency to stick with their own, there was, as at Sloy, little serious conflict between groups throughout the NOSHEB schemes. There wasn't even any resentment against the Irish who had been neutrals in the fight against Hitler. In fact, the Irish and the Highlanders generally got on well, perhaps because their rural backgrounds had so much in common.

However, relations between Irish workers and Lowland Scots could be very different. Scots from the industrial towns and cities of the central belt brought over a century of prejudice against Irish migrant workers with them to the schemes. Patrick Campbell, a native of Donegal, worked on the NOSHEB site at Dalcroy, near Pitlochry. In a vivid and thoughtful memoir of his days as a Tunnel Tiger he describes the shock and offence he suffered from unrelenting bigotry at the hands of urban Lowlanders. In camps further north though, the Irish aroused little animosity.

Don't imagine a politically correct Utopia. There was plenty of rough humour to be had out of national differences but it was essentially good-natured. Despite their nickname, the Bog-Arabs did not suffer discrimination nor was there any fighting between Catholics and Protestants of any nationality. Gaelic speakers newly-arrived from Lewis were jokingly turned away with the statement: 'Nae DPs in this hut!'. Individuals had disagreements with other individuals and indeed the Poles and the Irish had a reputation for quarrelling. But in all the interviews I had with men who had worked on the NOSHEB schemes, no mention was ever made of institutionalised strife between national or religious groups. What does emerge is a sense of the respect among the labour force for the courage and skill of the men undertaking 'the really dangerous work in the tunnels' and elsewhere. The workforce had been, I was frequently assured, made up from 'the kind of men who were prepared to tackle anything'.

Douglas Watson, who spent the early part of his career in civil engineering, gave me a very interesting perspective on the way the workforce had operated. It may well also explain the successful melding together of these different national groups within the workforce. 'The GFs knew their men, they were not socially remote from them', Douglas told me. He went on to explain that there was no social distinction made between the labour bosses and their rank and file. Everyone ate and lived together; the contractors' camps housed a strictly single-tier society. Even if the GFs were earning more, they were not seen as being essentially different from the men they were in charge of. They directed their men from alongside, not from above. A Dingwall tunneller confirmed this when he told me that 'a tunnel boss would never ask you to do anything he wouldn't do himself'.

The workforce undertook all the basic tasks of a civil engineering project plus the specialities of hydro construction: dams, cofferdams and tunnels. At the start of the Board's Development Plan, 60% more manpower was required than at its end when significant advances had been made in mechanisation, particularly for earth moving machinery. By 1961, hydraulics had replaced ropes on these machines thus greatly increasing their efficiency. The first job fifteen year old Alec Ross got on the Affric scheme was as chainman assisting the engineers who were mapping out the course of the tunnel from Mullardoch to Glen Affric. The job title derived from the lengths of chain which were used to mark out the engineers' measurements on the hillside. 'Little did I realise', he admits, 'what I was taking part in.' He was none the wiser when the engineer congratulated him: 'Well done Alec, lad: you've just opened the Mullardoch tunnel!' The building of the huge concrete dams involved skilled and carefully co-ordinated working practices that did not benefit greatly from any advances in mechanisation. The concrete was poured into wooden shuttering built from engineers' blueprints in the carpentry workshop and then brought to the dam site on low loaders. Concrete, already mixed, was delivered in skips that opened directly into the shuttering. Alec remembers these skips being called bounty wagons. Inspectors working for the Consulting (Civil) Engineers were responsible for the quality of the concrete. The men did not always take these inspectors very seriously. When asked, rather vaguely, by an inspector if he thought the concrete was 'alright', one Black Isle man could not help but recall how recently the same inspector had been employed as an insurance salesman.

The inspectors also had to look out for unsatisfactory working practices during the construction of dams, such as bagging. This occurred when concrete was rubbed with sacking to roughen and so conceal the surface of a badly made joint between the shuttered sections of the dam.

The scale of the dam-building operations impressed many of those who took part in them. Folk were struck by the vast quantities of concrete used, as indicated on many schemes by the full-time employment of one man who would spend his whole working day just opening cement bags.

The tunnelling process was a sequence of basic operations: 'It's an A-B-C-D job; you drill, you fire, you muck'. If all was going well, this sequence could be performed five or six times during an eight hour shift. Holes were drilled in the face in a spiral pattern into which gelignite was inserted. The gelignite was detonated in a delayed sequence, which caused the rock face to disintegrate and then collapse as debris on the tunnel floor.

Until the 1940s, the debris would have been removed to the tunnel mouth in wheelbarrows. But the Board's contractors used EIMCOs, mechanised shovels designed in America and manufactured under license in Newcastle-on-Tyne. These machines, referred to by the men as 'emkos', were used for shifting the spoil created as the tunnel progressed. Powered by compressed air, the EIMCOs collected all the rock dislodged by blasting and transferred it away from the face into small wagons which were pulled on rails to the tunnel mouth by locomotives. Most of these locos were powered by diesel and the sickly smell of diesel never left the tunnels. Men used to ride on the locomotives to get to the tunnelling face, saving a walk of sometimes three or four miles. Tunnels were not good places for taking risks; hitching this sort of a lift could have fatal consequences if the unprotected passenger hit anything protruding from the tunnel roof like a ventilation duct.

The tunnels were often half-full of water and gellie-reek also permanently filled the air. These fumes, given off by the explosive, had a sweet, sickly smell that gave the men headaches. The only lighting in the tunnels came from light bulbs screwed into temporary wires and powered by generators on the surface. Drilling created clouds of choking dust as well as a terrific noise and after a round of explosives had been detonated, the acrid smoke would reduce visibility in the tunnel to zero.

Tunnels were dark, hostile and frightening places. I spoke to a couple of men who admitted to having lost their nerve for tunnelling, one after coming to work one morning to catch sight of three corpses from the night shift waiting to be removed. Hugh McCorriston started at Cannich in 1947 and soon got used to working in tunnels. His brother joined him, saw one man killed in a tunnel shortly after his arrival and was so upset that he had to go straight back to Ireland.

But the tunnels did have some advantages. In winter they were out of the weather and in the summer they were free of midges. Midges could actually prevent evening shifts from working outside. A team overcome by midges on Loch Droma, near Ullapool, tried using compressed air to get rid of the insects, but to no avail.

The spoil removed from the tunnel was often used by the contractor's men for building new roads to replace any which had been flooded when the level of a loch had been raised. Even half-a-century later these dark sterile patches of rock spoil are easy to spot near hydro installations.

This outline of the labourers' tasks should not leave anyone in any doubt about the strenuous physical demands made by any one of them. More than once I was told, not with any arrogance, that today's workforce would not be equal to the job. In the 1950s, the war still exerted an influence. As can be seen in NOSHEB photographs from the early 1950s, many of the Development Plan labourers worked in army surplus garments. But the war also meant that men were used to pushing themselves physically, to 'getting their heads down'. They were certainly fitter than their car-owning descendants. Their working practices produced a highly resilient work force. Moreover, there was a powerful disincentive to complain: 'If you didn't like it you were out'. The rush to earn bonuses meant nobody had the time or inclination to complain about working conditions.

Health and Safety

You got the feeling that life was pretty cheap
Roy Mackenzie

The physical condition of the NoSHEB contractors' labour force is only one of the huge differences between the workforces' experience during the Development Plan and modern working conditions. There was nothing like today's complicated and bureaucratic employment procedure on the Development Plan construction sites. 'You'd be given some work to do in the morning and if you did alright, that was you taken on.' No-one was asked for qualifications when they arrived looking for a 'start-up'. Newcomers were assessed on their ability and performance and it just took the General Foreman a morning to find out all he needed to know. The only paper qualification I ever heard mentioned by former scheme workers was a licence to handle explosives. However, the licence itself did not signify any particular competence as one man who had worked in Glen Affric and at Invermoriston told me: 'some men were handling explosives who shouldn't have been'. I asked how men got one of these licences. 'Och, a bobby asked ye a few questions and some of them [the police] didn't know much about it'.

Sloppy explosive practices were not just a danger to tunnellers. A young civil engineer, right at the start of his career, was supervising the design and construction of the Mullardoch dam. He was watching an explosives squad at work at the far end of the loch. When he heard the drilling stop, he knew the charges were all set and that a remotely triggered blast would soon follow. On his way to find a sheltered position, he suddenly caught sight of wires all set for detonation on the ground in front of him. 'It was so close I could have touched it; I was paralysed with fear, totally petrified. You know, there's a split second when you know something must be done but you can't make a muscle move. Finally, it seemed, I did move, fast enough to get away, but not to get beyond [the blast]. But I got just far enough; some bits of rock went over my head.'

This lack of systematically scrutinised control characterised all work on the NOSHEB construction schemes. The contractors' use of bonuses to reward good progress rates meant that General Foremen, who stood to gain the most from bonus payments, tended to concentrate on getting work done quickly rather than carefully. A prime example of this dangerous haste was the failure of tunnellers to scale tunnels properly as it slowed down progress. Scaling is the removal of loose rock before drilling and omitting proper scaling greatly increases the danger of accidental rockfall. Even in the late 1950s men were only protected against this hazard by hats made from compressed cardboard.

Another potentially fatal short-cut practised in the tunnels was the men's reluctance to use water to damp down the clouds of rock dust produced by drilling. It was simply quicker to proceed without this safety measure; the resulting damage done to the workers' lungs could have been immense.

The contractors did not allow trades unions on the schemes at all and without health and safety legislation there was nothing to prevent this headlong rush for personal gain. Intent on their bonuses, the workers were quite happy: 'Unions was the last thing we wanted. They only would have held us back'. The contractors did have one single safety rule, however, and it was enforced rigorously. Drinking alcohol was absolutely forbidden during working hours. Anyone hoping to be taken on would be turned away immediately if he had drink on his breath and being discovered drunk at work meant instant dismissal.

However, the contractors' determination to finish the job in the shortest possible time meant that they encouraged the men to work long periods of overtime; a doubler lasted all day and all night and a ghoster was a doubler plus another twelve hours. Such long shifts were just as dangerous for the men as drinking on site. Many have testified to their effects. After a ghoster spent tunnelling in Glenstrathfarrar, one Dingwall man told me, 'I was shattered. I couldn't walk straight. I couldn't even think'. The combination of long, long shifts, with bonuses and a total lack of any statutory legislation to protect the workforce meant that accidents were commonplace throughout the schemes.

The NoSHEB construction schemes in Glenmoriston had a particularly high accident rate although no one who worked there, not even James Black, can explain it. I was lucky enough to talk to Roy Mackenzie, originally of Fortrose, who had joined the police after serving with the Royal Marines in World War Two. He was the local policeman at Invermoriston while the NoSHEB schemes were under construction there. His admirably detailed memories provided me with a comprehensive catalogue of the accidents that happened there as well as an insight into the operational culture that allowed so many injuries and fatalities to occur. The absence of any state regulation of health and safety at work meant there were no independent inspectors to monitor accidents and decide on their causes. It was the police who had to record officially what had happened when there was an accident. Mackenzie was able to give me a very clear picture of the way the bonus system undermined workers' safety, having reported on hundreds of accidents during his time at Invermoriston. Among the men at Cluanie Camp, which at its busiest housed 1,000 workers, there were 22 fatalities in two and a half years. All 22 deaths resulted from accidents sustained at work apart from one when a man died from a heart attack while being chased down a tunnel by a Pole.

According to Roy Mackenzie, all the accidents he saw could have been prevented. He believes they were 'the result of the workers pushing on and disregarding safety'. Explosives, dangerous in the most competent hands, were a frequent cause of injury and death. Roy had to report on one accident that happened four miles inside a tunnel. The accident had occurred when gelignite had ignited prematurely. The blast killed one man outright, though miraculously the rest of the charges waiting to ignite did not explode. But the body of the dead man had been struck so powerfully that 'we never found his left hand yet'. The search must have been particularly difficult, as the tunnel was partially flooded.

'Nearly every tunnel had a fatality', as I was told during my conversations with former NOSHEB labourers and I certainly heard about a few of them. A Strathpeffer tunneller told me how a group of workers was caught by an explosive charge while they were having a tea break underground. No one had realised they were in range when a set of charges was exploded and the entire group was killed.

The youngster, Alec Ross, saw an explosives fatality in Glen Cannich at very close quarters. In the confusion of a severe winter storm, a man had made the classic mistake of stuffing in more gelignite with a ranging rod (a six foot metal measure used by engineers for surveying) where a charge had failed to detonate. Alec was given an ominous instruction by the GF: 'Go and see if Paddy's alright; I don't think so'. Indeed he wasn't and Alec was given the job of burying the dead man's stomach. Back at work the next day, comments from the older men were all he got by way of counselling: 'It was about time you grew up, boy'.

Above ground, where big lorries and earthmovers were constantly on the go, conditions were also hazardous. One man died at Invermoriston when he was at the tip end using a tipper lorry to empty spoil from the tunnel. (The tip end is the place, above ground where spoil from the tunnel is deposited.) One of the lorry's axles cracked, the driver lost control of the vehicle and it went over the edge, killing him. A piece of very careless driving cost another five lives above ground. A lorry was being reversed downhill with the driver standing on the running board. His foot slipped off the brake and the lorry plunged backwards down the hill, smashing into a tea hut at the bottom and killing the five men who were inside at the time.

One accident in particular, which I learnt about from Roy Mackenzie, showed just how hopelessly inadequate the working practices of the NOSHEB contractors were, especially in the dangerous environment of the tunnels. The workers involved in the accident were engaged in driving a surge shaft upward off the main tunnel, known to be a risky process. As one man from Dingwall told me: 'We had to do this now and again and it was dangerous because it [the roof] might fall in on you at any moment'. Surge shafts had to be built to cope with the build up of pressure which results after water is halted suddenly in a tunnel when, for example, generating operations are shut down quickly.

A man was sent onto a 300 foot high gantry in the shaft, 'to see if the gas [carbon monoxide] had cleared' after blasting. This strategy could be compared with asking someone to find out how busy a motorway is by telling him to go and stand in the middle of it. Sure enough, the man was gassed and he dropped to his death from the gantry. Another man was sent to investigate and when he failed to report back, a third had to follow him. He found the second man overcome by fumes. He tried to lower the weakened man down on a rope but he didn't tie it properly and the second man fell to his death. The third unfortunate was so shocked by what had happened to the second that he overbalanced and fell off the gantry as well. All three men, Roy told me, were from the north-east and left 21 children between them. There was no question of compensation being paid to the families of anyone injured on the schemes and after that accident Roy told me that he had realised 'these hydro schemes were being run for profit not safety'.

Certainly, the men preferred to make the minimum of fuss if they were involved in an accident. One man let his hand get in between the couplings of two loco wagons. 'Ach, Jesus, there's my thumb', was his only comment and, after getting his stump bandaged, he went on and finished his shift. If a man were killed at work, the men would sometimes down tools out of respect. But shifts were also completed after disasters had occurred: 'People were used to seeing fatalities,' a veteran NoSHEB tunneller told me, 'and because things were done in a hurry, accidents happened'.

Why did the men stay on working in such perilous conditions without even the guarantee of compensation if disaster struck? The answer lies with the fabulous wage earning opportunities presented by the NoSHEB Development Plan.

Wages

There he was lying in the bath: dead drunk with fivers and tenners floating on the water... There must have been hundreds of pounds in there...

Until the start of the NoSHEB project, the main source of work in the Highlands was the area's private estates. For some this meant agricultural, gardening or gamekeeping work while others, mainly women, performed domestic duties. There were some non-cash advantages to be had from this sort of employment, like the provision of accommodation, but the hours were long and wages were low. Even after the Second World War, workers on Highland estates were paid a maximum of 2/- (10p) per hour. There was no question of bonuses or overtime.

Contractors on the NOSHEB construction sites were paying at least double that amount and, because of the contractors' determination to maximise progress, the men could do as much overtime as they wanted. No wonder that Highland landowners tended to disapprove of the NOSHEB and all its works. The lairds' Highland lifestyle, even if it only involved a few weeks' residence each year for stalking or fishing, depended on a permanent supply of cheap labour. The hydro schemes provided Highlanders with an opportunity to earn more than this hard-won pittance without having to emigrate. The Forestry Commission offered a few Highlanders an alternative to estate work but even a relatively senior forestry worker couldn't expect to take home more than £10 per week while the boys in the tunnels could be making at least £30-35 per week. Hugh McCorriston told me that his father had earned over £60 per week as a tunnel boss at Mullardoch. Before he decided to follow his dad to the Highlands, he had been earning £3.50 per week in a timber yard back in Northern Ireland. Roy Mackenzie, the Invermoriston policeman, as he broke up the workers' drunken fights and attended harrowing accident scenes, would have known that the tunnellers were earning over ten times more than his weekly £3.50. It is a measure of his humane intelligence that I could detect no trace of retrospective resentment in his tone when this comparison of wages was discussed: 'They worked hard for their money, they really did'.

So the hydro boys made the most of the golden opportunities offered by the NOSHEB Development Plan. The absence of trades unions and the dangerous working conditions would not keep them away from such a bonanza, Highlanders had never had the chance to earn this kind of cash before. Alec Ross was earning £3 per week working for an undertaker in Inverness when his father, Ebenezer, took him to Glen Affric.

Eb, who drove workers from Inverness down Loch Ness-side to the Cannich Camp everyday for Paterson's of Strathglass, showed Alec the bustling activity of men and machines at Cannich and told his son to forget about his job in Inverness: 'There's plenty money down there, boy, and we're going to get ours'.

The sudden possibility of huge cash earnings led to mighty and continuing celebrations. These celebrations would carry on throughout the years of the Development Plan and we will hear about some of them in the next section.

Camp Life

The place was going like a fair.

Alec Ross listened to his dad and went to Cannich every day on Eb's bus until he left to do his National Service with the Royal Engineers. In fact, most men simply stayed in the camps set up by the contractors. At that time, the condition of Highland roads made them harder to navigate than is the case today. Even for men coming, say, to Cannich from the Black Isle, it was more practical to live on site and only return home at the weekend. Few people owned cars and most of the workforce stayed in the contractors' camps for months at a time.

NOSHEB contractors were responsible for the setting up and running of the camps where the majority of the workforce stayed. They or their sub-contractors provided the men with accommodation, food, medical care and, in some cases, entertainment. Some camps had better reputations than others. Less scrupulous contractors, trying to save money would spend as little as possible on food and general facilities. The camps all had a military appearance with ranks of wooden huts, each one housing twenty men. The huts all had a stove, bunk beds and lockers for the men's personal possessions.

The men paid a weekly amount of around thirty shillings for their accommodation. This covered their hut, stamps for meals in the canteen, a daily packed lunch and cigarettes for seven days. The official ration was ten cigarettes per day, 'but if you were nice to the girls they might give you twenty'. Payment of this weekly sub entitled men to use the camp's medical facilities.

The huts were grouped around a central, larger building where the men ate and where entertainments could be staged. Hugh McCorriston met his wife-to-be, Amelia, a local forestry worker's daughter, at a dance in this main building at Cannich camp. Cannich has used this hut as a village hall ever since the workers left and Hugh and Amelia are its caretakers.

Most camps were completely dismantled when the job was over. I've been shown deserted spots all over the Highlands and heard stories of the time when the same place was 'going like a fair'. Trees have grown up and only the odd clue, like the stone wall of an old diesel tank housing, remains of the activity of half a century ago.

Although the camps looked very similar, their reputations varied considerably. The Muir of Ord contractor, Willie Logan, always had his wife in charge of the camp and the catering, and the workforce enjoyed the results. Other contractors put on poor food, 'so bad you could smell it was bad'. Students took vacation jobs on the schemes and they were often given work preparing the men's food to save their hands from the rigours of hard manual labour. The men approved; at least the students' hands were clean.

Cost-cutting contractors also resorted to the practice of 'hot bedding'. This meant that the night and day shifts shared the same beds which were never allowed to air. Many believed this practice caused infestation and illness. Some contractors only supplied blankets and then the men often slept fully dressed. In some camps they had to keep their boots on as well to stop them being stolen or used as impromptu urinals by men who didn't want to go outside in the cold.

So contractors' standards certainly shaped the workers' experience of a camp. But enjoying camp life also had a lot to do with 'getting into a good hut'. 'North men' would try and get into a hut together and Highlanders also considered the Irish to be 'good crack'. A 'good hut' often held its own ceilidhs, the men using accordions and fiddles which they had brought with them from home. The Scots and Irish knew many tunes in common and the Irish would stepdance them 'to perfection'. While these hut ceilidhs were highly enjoyable for the participants, not everyone appreciated them. James Black remembers an ex-Irish Army Major at Sloy who burst into a late-night gathering and shot his revolver at the stovepipes to bring the party to a close.

Some contractors, like Cochrane's in Cannich and Bisset's at Grudie Bridge, put on live entertainment at the weekends. Top class musical theatre acts were brought up from Glasgow and films were also shown. Workers had the choice of two camp canteens: the dry canteen, which served food and tea, and the wet canteen, which served beer but not spirits. At Cannich, both canteens were housed in one building where they were separated by a corridor. If the camp was close to a pub or hotel, men would have to go there if they wanted a dram and that was where they congregated on a Saturday night if there wasn't a dance on locally. As we shall see, hydro workers usually mingled happily with locals at these dances. However, in the pubs and hotels the men from the schemes were only served in the Public Bar and never admitted to the Lounge where the engineers drank.

Inevitably, some men flush with their bonuses and away from home and family did a lot of drinking. If there was no pub nearby the men would borrow a contractor's bus, take off to the nearest town and drink as much as they possibly could. One former tunneller explained: 'There we were, stuck out in the back of beyond, 3,000 men with nothing to do but queue for beer. We had to do something'.

Alcohol caused wild behaviour and sometimes trouble, especially fighting, amongst the men. Roy Mackenzie told me that crime on the camps was 10% thieving and 90% fighting. He confirmed that drink fuelled this fighting on the camps and in local pubs. The sight of a knife being brandished was not unknown but drunken scrapping between workers was not normally of a deadly serious nature. In his five years at Invermoriston, Roy Mackenzie didn't have to deal with many serious assaults. Hugh McCorriston only recalled one nasty scrap at Cannich when two Poles had 'cleaned out' all the lads they'd been playing dice with and their victims wanted revenge.

'It was the roughest that was in it' as I was told, but the roughest seem to have been basically a decent lot. Even Dirty

Dick, who never washed, and painted the surveyors' guidelines on the tunnel roof with his jersey sleeve, didn't do much worse than the common tunnel trick of urinating into another man's wellie undercover of the tunnel gloom. But I have never heard a word about bullying or, as already mentioned, about endemic racial harassment. I asked a tunnel veteran how a workforce of such mixed origins had apparently got on so well together. His answer was simple: 'We might have belonged to different places, but we were all miners.'

There seems to have been a minority hard core of heavy drinkers who spent most of their wages on drink; I heard them described as 'long distance men: more or less hobos'. They didn't stay long on any scheme. One old drunk wound up as a loco driver at Invermoriston and died of a heart attack in the tunnel. Fortunately, he must have known very little about it, but everyone else soon realised what had happened when the loco and its dozen full wagons sailed over the tip end.

Recreations

I tell you, it was real life, it was really an experience.

Not every worker's leisure requirements were fully met by official camp activities and unofficial ones were rife. Gambling was ubiquitous on all the NOSHEB schemes including Sloy, though it had to take place under cover. The men gambled amongst themselves as they had done at Kinlochleven and for the same reason: boredom. But the vast amount of money that the NOSHEB boys were earning gave their games an extravagant recklessness. It was not unknown for men to wager their pay packet, unopened.

Professional gamblers (it is often suggested that they came from Glasgow) came to the camps, usually when the men had just been paid. They would slip away again on Monday mornings carrying suitcases stuffed with NOSHEB cash. One of these professionals, known as Gambler Docherty, took a Crown and Anchor board round the camps and was always willing to lend a stake to hard-up men against their next week's wages. Docherty always carried away his winnings, often as much as a thousand pounds, in a small leather suitcase. One night he was followed into town by men from the camp he had just visited. His body was found later that night on a local bus but the money was missing. It turned out he had suspected foul play and given the case to a local woman to look after. No-one was ever charged with his murder.

The police tried to limit gambling to prevent the arguments caused by heavy losses. There was a police presence on all the camps. As well as dealing with accidents and maintaining order, the police scanned the workforce for anyone featured in the weekly *Police Gazette*, 'to keep out the wrong 'uns' as Roy Mackenzie told me. One Irishman who had worked building the dam at Lairg, had been most impressed by the camp policemen there: 'They were big, tough Highlanders who kept things well under control and if not, you'd get a couple of thumps on the chin round the back of the hut'.

Gamblers were not the only ones to visit the camps; women also smuggled themselves in at weekends. Huts were requisitioned as 'hen-houses' and the women's services offered and discreetly paid for. The men from Cannich sometimes took the Saturday bus into Inverness where the general aim was to enjoy the Highland capital's hostelries. Clem Watson remembers, as a young boy, seeing the streets of Inverness on a Saturday afternoon thronging with St Patrick's Day revellers from the hydro schemes. Hugh McCorriston often came back to Cannich camp from Inverness on a Saturday evening with quite a few of the lads he had travelled into town with in the morning missing from the return bus. 'They'd have been arrested for Drunk and Disorderly and kept for the night at the Police Station,' he explained. 'The Bobbies would fine them whatever money they had in their pockets, leaving them £1 for their bus fare back to Cannich.'

Saturdays in Lairg for the workers on the remote Shin scheme were rather more predictable: a cup of tea, a haircut and a few pints at the Grange Bar was the routine for most. The men never used the grander Sutherland Arms at all.

Shopping was another weekend activity; some workers travelled to Inverness, Oban or Fort William every fortnight. They would buy a new suit, shirt and underwear, change into them in the shop and wear the lot until their return to town a fortnight later when they would repeat the process. The Irish were well known for tunnelling in suits. Eventually, itinerant traders realised the value of the captive market stuck out on the hydro schemes with plenty of cash but nothing to spend it on. Asian peddlers brought dungarees and vans arrived selling all sorts of goods and provisions, though the catering contractors protested loudly if the monopoly of their camp shop was threatened.

The contractors' men, especially the tunnellers, were earning such large amounts of cash that on these weekend shopping expeditions they regularly reached new heights of drunken excess. One Dingwall tunneller walked from Inverness to Loch Awe without a penny in his pocket looking for a 'start-up'. He and his mates ate what the tourists had left in the rubbish bins and slept overnight in roadside fields normally the preserve of cattle. Individual cows were forced to shift so the hydro boys could take advantage of a ready-heated patch of ground. At Cruachan, the men were recognised and taken on immediately on the strength of their tunnelling reputation. They were lent money to keep them going until their first week's wages: £200 in cash.

They were soon treating Oban to the spectacle of 'the weekend millionaires' in brand new made-to-measure suits, wellie boots and cravats. One gang of prodigals spent £350 on a brand new red mini. On the way back to camp they hit another car. No one was hurt but the mini didn't stop. Its occupants soon realised their mistake but what should they do? They had to get rid of the incriminating vehicle, so they filled it up with stones and pushed it into a loch. When the police visited the camp looking for the boys with the red mini, action had to be taken. What simpler than to go to Glasgow and buy another red mini to throw the police off the scent? 'Wasn't that rather a lot to spend in one fortnight?' I asked a Dingwall tunneller who was in on the story. 'We didn't care: it wasn't our money. There would be plenty more the next week.'

All in all, the hydro boys indulged in quite a few nefarious pursuits. The rural setting of the NOSHEB schemes made poaching for salmon and venison a common pastime. Even a very old man who had run the Ross-shire Electric Supply Company power station at Loch Luichart before the war told me that he had plenty of time during his working day to pop out and take the odd rabbit. A

Ross-shire man at Invermoriston chose to rent a cottage with some pals from the north coast rather than stay on the camp. This was so that the four of them could pursue their salmon-netting interests unnoticed. They were lucky and regularly sold their catch at the Market Bar in Inverness.

One particularly enterprising poaching operation used the efficiency of the Board's state of the art salmon management technology. To satisfy the fishing lobby, the riparian owners' hereditary rights to all the returning salmon passing through their property had to be respected utterly. Ingenious fish passes were built into dams on rivers to help salmon find their way upstream. Fish-counting equipment was sometimes incorporated into the fish passes, enabling the Board to provide the owners with the sort of statistics highly prized by modern game management. Fish passes also had something else to offer though, but not necessarily to gamekeepers.

When a dam was being built on the lower section of the Affric-Beauly Scheme, one enterprising construction worker made a copy of the fish pass key. When the dam was in operation, he could return, activate the pass and simply collect all the salmon waiting to be guided through the dam.

Pilfering was widespread on all the hydro schemes. The quantity of stores unprotected by the precise supervision of modern cost accounting was irresistible to folk who could profitably dispose of them. The process was not always a subtle one; Alec Ross remembers returning to work one Monday morning at Cannich to see a whole set of farm buildings freshly decorated with the red lead paint the contractors normally used to protect exposed metal work against corrosion. Explosive charges were perfect for dynamiting fish and tons of scrap metal was there for the taking, 'if only you could work out a safe time to uplift it, given the 24 hour shifts in operation'.

Even old Eb Ross had a scam going which capitalised on the workers' love of gambling. There was usually a card school in operation on the bus home to Inverness in the evenings. Eb would brake sharply toward the end of the journey, scattering cards and coins all over the floor of the bus. He would be sure to find some money left on the floor after the men had gone.

The Locals

I see them in my sleep.

Wherever the NOSHEB had its construction schemes, the labour force and the locals met at the pub and at dances in the village hall and generally got on well together. Some locals worked for the contractors and many of the NOSHEB labour force came from rural communities in Scotland and Ireland that were very similar to the places where the schemes were under construction. Dances were held regularly in the village hall at Garve and the men from the Fannich and Grudie Bridge camps would come along. The hotel there refused entry to the men from the schemes.

In Glenmoriston too, the local community and the hydro boys were well integrated. Roy Mackenzie told me that whenever a serious accident or a fatality occurred on the construction scheme, 'the whole village would be affected'. There seems to have been little outright conflict between the locals and the temporary incomers anywhere throughout the NOSHEB construction project.

However, remote Highland communities had very little experience of strangers in such vast numbers. Throughout the Second World War, much of the area north of Inverness had been a Restricted Area where no visitors were allowed without official permission. Not surprisingly, therefore, there had been some apprehension before the schemes started about the disruption the NOSHEB influx might cause.

Many of the older folk expected a visitation that must be their unfortunate due because they had escaped the worst nightmares of blitzkrieg or invasion. They worried that the incoming workforce would threaten the peace of their villages with rowdiness and lax behaviour. There was also genuine anxiety about the potential influx of large numbers of Irish Catholics. This was not just casual prejudice; many Highlanders were being told from the pulpit that Catholics were heathens who would burn in hell. The prospect of intermarriage was, therefore, quite appalling. For all the communities where schemes were planned, fear of the unknown was very powerful.

But people's worst fears were not realised. At Lairg, 'even though they'd laid on two extra Bobbies', there was no crime wave.

The worst of the fighting there was usually started by a local man who 'liked nothing better than taking on the hydro boys'.

In fact, the NoSHEB Development Plan brought plenty of incidental benefits to Highland communities, not least 'a bit of excitement and something to talk about' in places where there had been few significant arrivals since the appearance of the Lowland shepherds at the time of the Clearances; the Hydro Board and what it was getting up to 'was the topic'!

Communities certainly enjoyed access to the facilities laid on by the contractors for the workforce; locals were able to go to films and concerts put on at the camps. One lady I spoke to heard an electric guitar for the first time at a show put on at Fannich Camp. The locals were also able to expand their limited shopping opportunities by using the camp shops. In several places the decline in school rolls was reversed by the arrival of families who had accompanied men, usually engineers, to the construction schemes. A lady from Orkney, who worked as a secretary for the engineers, played the kirk organ 'beautifully' at Lochluichart; after she left, the Minister was reduced to using a tuning fork.

What had it been like for the locals while the schemes were under construction in places they had been used to having to themselves? Brother and sister, Willie and Jessie, grew up on the Lochluichart Estate where their father was the head gamekeeper. They have an enduring love of their birthplace and very kindly shared their memories of the Development Plan era with me.

Their father's key position on the Lochluichart Estate meant he was involved in frequent negotiations with the contractors over their activities which affected estate land. He had to deal with tenants' complaints about the noise and disruption of blasting as well as keeping an eye on any damage done to estate land by the contractors. Willie told me that his father had found these unexpected responsibilities taxing: 'He was near the end of his working life and these new duties caused him plenty of problems'. Many of the older Highland residents found the activities of the Board's contractors similarly disconcerting. Suddenly there were gangs of labourers everywhere and even the landscape was being altered. At Lochluichart, the creation of a reservoir enlarged the loch. This meant the flooding of several farmhouses, a walled garden, the railway station and two railway cottages. For the first time on the British mainland, the railway line (from Inverness to the Kyle of Lochalsh) had to be re-routed and a new station was built at Lochluichart. Young and old locals alike had suddenly to get used to the presence of thousands of strangers where previously there had been only familiar faces plus the occasional visitor.

Jessie, who travelled to work in Dingwall every day, soon found herself missing the peace and quiet which the presence of the contractors and their men had banished from the neighbourhood. Her evening walk home from the station became something of a challenge, especially in the winter darkness. She told me that, although nothing ever happened to her at the hands of the hydro boys, 'You always felt the fear, which wasn't very nice'. She and the other locals went from knowing everyone in their quiet, remote community, to a very different situation where suddenly 'you had got a lot of strangers'. Jessie was surprised to find a television crew preparing to film the opening of the new station as she set off to work one morning. Nobody saw the piece in the Highlands: television did not arrive in the area until the 1960s.

Jessie and Willie continued to enjoy dances at Garve Village Hall, now well attended by men from the camps. These were relaxed and pleasant occasions. Jessie and Willie agreed that overall, the locals and the hydro workforce 'got on really well together'. I asked if locals who were still working in traditional jobs didn't resent the enormous wages being earned by the men on the schemes. There were no hard feelings, Jessie and Willie told me, 'because the locals knew how dangerous much of hydro work was and they weren't wanting to do it themselves'.

People all over the Highlands found the sudden appearance of troops of workmen unsettling, to say the least. 'You never knew when you would come across them, cutting down trees or making some sort of disruption,' as one man from Lairg remembers. At Lairg, fleets of yellow lorries carting cement were on the road 24 hours a day and many locals dreaded them: 'I see them in my sleep' one man complained at the time. The dangers of hydro work scarcely affected the locals at all. Jessie remembers hearing the sound of blasting from her bedroom as the night shift got to work building a tunnel in the hillside that would bring water down to the turbines at nearby Mossford Power Station. I wondered if this had made her want to go and have a look for herself at what was going on underground. 'Women were not allowed in the tunnels,' she told me. 'It was thought to bring bad luck.'

Although no actual harm ever came to herself and her family during the construction period, Jessie and Willie did suffer one family disaster at the hands of the Hydro Board contractors. Connie, the Kerry cow, a vital component of the family's domestic economy, died after straying onto the site where Mossford Power Station was being built. Jessie is still convinced that the contractors' red lead paint killed Connie, 'but we never got a penny of compensation, not a penny'.

Naturally, there were those in the Highlands who welcomed this great frenzy of activity. Businesses, not least pubs, benefited dramatically. One hotel close to a major camp is reputed to have financed the addition of an entire new storey thanks to the money it made from the contractors' men. Crofters like Jessie and Willie's family sold eggs, milk and cream to the local camp caterers and travelling traders visited the remote schemes. And of course, the children just loved all the excitement.

In the peace which reigned across the Highlands before television, the clatter and commotion of the civil engineering works was constantly entertaining for youngsters. Jessie and Willie were so astonished by the sight of huge quarrying equipment when it arrived at Lochluichart that they both came off their bikes. A primary school pupil while the Board was developing Loch Shin, Willie Ross remembers the tremendous excitement he had felt, watching the scheme from start to finish. The first sign of activity that he noticed was the appearance of surveyor's yellow pins marking out the projected level of the new reservoir on the hillside below his family's croft on the western shore of Loch Shin. From then until the completion of the scheme, Willie and his friends were enthralled. As they made their daily journeys to and from school, they would always be sure to check 'how the hydro boys were getting on'. Back home, Willie kept a close eye on developments using his father's telescope to scan the changing landscape.

Boys at Lairg, Willie Ross remembers, made friends with some of the workmen, especially the Irish. 'The Bog-Arabs were quite friendly: they got us boys to write their letters home for them', he recalled. Indeed there is a general recollection that some of the Irishmen employed by the contractors were virtually illiterate.

Douglas Watson, another primary school boy who watched the enactment of the NOSHEB Development Plan, remembers the almost shocking way that the Board's contractors filled Glenmoriston with incessant noise and activity. He was also fascinated by the filthy appearance of the workforce. Before the start of the NOSHEB Constructional Schemes, crofters, teachers, shopkeepers and the forestry workers who worked with his father were the only adults he was used to seeing. Unlike the hydro boys, these adults were never dirty for long. He was also aware that the hydro boys were regularly involved in 'plenty of wild nights' in the local pub. He thinks it more than likely that this vivid presence in his childhood influenced his eventual choice of a career in civil engineering.

Willie Ross had also been struck by the grimy appearance of the hydro workforce; when he saw queues of men waiting patiently to visit a prostitute in Lairg, he was struck by how 'dusty' they looked.

The children and indeed the whole community were always involved in the celebrations staged by the Board at the start and or completion of every construction scheme. Tom Johnston stayed on as chairman of the NOSHEB until virtually the end of the Development Plan's execution. He remained the master publicist and never overlooked a single opportunity to celebrate the Board's efforts in the communities they had been designed to benefit. I found out more about this aspect of the Board's activities from Hamish Mackinven. He told me that Johnston made such celebrations compulsory for 'even the tiniest, loneliest distribution scheme.' In his time, Mackinven had to organise a variety of such occasions. Some were gigantic PR operations like the celebration of Iona's connection to the Grid in 1958. Iona's iconic importance in Scotland meant that its connection had to be marked with the maximum of pomp and ceremony. Mackinven was told to charter MacBrayne's largest steamer, *George V*. This renowned vessel was to carry to Iona the crowd of distinguished guests and journalists who had been brought to Oban, first class, by train from Edinburgh and Glasgow. At the end of a day of speeches and switch-ons, with everyone back on board the steamer for the return trip to Oban, Johnston asked Mackinven to send in Bob Brown, then Scottish Correspondent of the London *Times*. The chairman knew that by giving Brown an exclusive interview, the NOSHEB's triumph at Iona and throughout the north would get at least two columns in *The Times* the next day.

Willie Ross remembers very vividly being the beneficiary of the NOSHEB's less elaborate partying when work on its scheme at Lairg began in 1953. All the children were given a half-day holiday from school and a poke of sweeties while the grown-ups got a dram. Everybody could have a good look at the visiting dignitaries imported by the Board to invest the proceedings with suitable gravity. The grandest of all at such occasions was undoubtedly Queen Elizabeth, the late Queen Mother, who opened the flagship scheme at Sloy in 1951. (I did hear about a GF's secret and vital role in that particular ceremony. Out of sight, he turned on a concealed pump when the Queen ceremonially activated the switch that would start the scheme's operation. This pump had to produce agitation in the tailrace because the turbines couldn't do so. Damaged during testing, the turbines had been out of action for months and there had not been time to fix them before the official opening.) In Lairg, in 1953, the job of ceremonially opening construction operations went to one of the village's oldest lady residents. She activated a remote detonation which began the excavation of the Lairg dam's foundations.

Willie from Lochluichart went to work for the Hydro Board in 1951 after his National Service. He helped set up the seats and bunting for these opening ceremonies at Lairg. Tom Johnston made a speech promising to find a job for any unemployed man present. Whenever the chairman appeared at these formal occasions, he impressed ordinary people with his obvious concern for their welfare.

However by 1959, it was obvious that the Board's prime aim, the attraction of major industrial players to the Highlands had failed. In fact, emigration from the area had not been reversed. Electricity had made life easier and more comfortable for a lot of folk who, without the Board, would never have been able to afford its advantages. But in many other ways the standard of living in the Highlands still lagged behind the rest of Scotland and Britain. New employment opportunities simply had not materialised in significant numbers and so wages remained depressed. Many Highlanders who had been away in the war now had first hand experience of the relative comfort and ease prevailing in other parts of Britain. They would also discover that electricity could not compensate for the absence of employment opportunities. Very little in the way of industrialisation or economic development had occurred to change the region's poverty in the 60 years since Colonel Blunt-Mackenzie had struggled to establish his supply industry in the face of an obstinately low demand.

Wingy Tam

To Thomas Johnston, electricity was just something you pressed and the light came on.
Hamish Mackinven

Johnston's dedication to the welfare of the Highlands, as evidenced by his work for the NOSHEB, is beyond question. But his failure to make sure the Board's efforts actually brought significant job opportunities to the Highlands was a disappointing outcome. The perennial difficulties of Highland economic development, especially transport costs and sparse internal markets defeated even the great man himself.

In the 1990s, the electrical infrastructure put in place by the NOSHEB made it possible for the Highlands to use information technology to mitigate the area's remoteness. The resulting opportunities for Highlanders have included all sorts of education and training opportunities, the most notable being the new university, the UHI Millennium Institute.

Part of the Highlanders' pessimism about the advent of the schemes was the certainty that any economic benefit they brought would only be short-lived: 'It won't last,' was the locals' common refrain even at the most hectic times. While people's fears about the moral threat of Roman Catholicism were never realised, their cynicism about a permanent increase in job opportunities was to prove wholly justified. On each of the schemes, the story was the same. By the time these opening ceremonies were being performed, the bulk of the labour force would have moved out, leaving behind a handful of NoSHEB employees in charge of running the power stations and keeping an eye on the dams, tunnels and other installations. Heavy industry stayed away, more or less; by 2001, former NoSHEB generating schemes all over the Highlands are monitored and managed by computer from a control room at Grampian House in Pitlochry.

I was lucky to be allowed a look at the operations room there where a dozen screens blink a constant flow of data 24 hours a day, every day. Framed black and white photographs showing hydro power scenes from yesteryear decorate the walls, including one of the Grampian Electricity Supply Company's power station at Rannoch. The station's staff is grouped in front of the building; it took about the same number of people to run one power station in the 1930s as the remote control of 56 power stations requires today.

One final paradox of the NoSHEB project concerns its questionable safety record. Tom Johnston was determined to have the Board work through its Development Plan as quickly and completely as possible to avoid attack from powerful enemies in and out of Parliament. The huge volume of work and the Board's haste to complete it meant that contractors were only too well aware of the need to impress the Board with their progress rates. These rates made excellent PR, helping individual contractors ensure they got a good share of the rest of the NoSHEB construction work. Johnston's publicity machine made sure everyone knew how substantial that work was going to be and so his haste transmitted itself, via the contractors' greed for future work, to the labour-force. As a result, these labourers were made vulnerable to accident and injury by the bonus system and the insanely long shifts in which overtime was done. The banning of trades unions by the Board's contractors meant the men were powerless to combat this state of affairs.

Johnston turned down a peerage when he retired from the Commons in 1945. He certainly cannot be accused of becoming a fat cat and betraying his principles. But for the scores of men killed and injured during the execution of the Development Plan, a question must be asked. Did Johnston, hero of the Scottish working man, sacrifice the lives and well-being of the labour force on the altar of Highland Water Power?

The labour force had a nickname for the Chairman: they

called him 'Wingy Tam' because of the withered arm he had had since childhood. But he held himself aloof from the men unless there happened to be a press photographer present. Only for the sake of the publicity machine would Johnston allow his remoteness to be breached.

The Board's laissez-faire attitude to labour welfare continued until the start of official health and safety legislation in the 1970s. I heard about one man who was seriously injured during the Board's redevelopment of Foyers in the late 1960s. He was the victim of the sort of carelessness which had typified the bonus-driven health and safety shambles of the earlier Development Plan. After months in hospital he abandoned the apparently hopeless struggle for compensation from the contractors. With a wife and young family he had no choice but to give up and get on.

Yet this paradox of a socially motivated employer operating without a perfectly-functioning social conscience cannot cancel out the worth of the Board's work in the Highlands. Perhaps it's anachronistic to criticise NOSHEB safety standards; all industrial production was potentially hazardous before the introduction of Health and Safety legislation in Britain in 1974 and is still dangerous today. Whatever the shortcomings and contradictions of the NOSHEB project, Tom Johnston made a supreme effort on behalf of the Highlands, for which he deserves to be remembered. Without his efforts the area would be unrecognisable today.

But there has been an important omission since the end of the Development Plan. There are memorials at some of the former NOSHEB schemes for workers killed during their construction. The most impressive of these is the memorial arch outside Clunie Power Station, Pitlochry. It bears the names of seven men who died in the tunnel there. The manner of their passing highlights the fatal risks of tunnel work as lightning, striking above ground caused charges set in the tunnel face to explode prematurely. The arch was made as an exact cross-sectional replica of the pressure tunnel which delivers water to the turbines in the power station and measures twenty-two feet in height, giving some idea of the massive effort involved in the tunnel's construction. But there is no monument anywhere in the Highlands to commemorate all the men who gave their skill, their strength and their lives to bring Neart nan Gleann – Power from the Glens.

Blackwater Dam under construction (1904–1909) (© Alex Gillespie)

Stone-crusher, Kinlochleven, 1906 (© Alex Gillespie)

Lifting gear for sluice gates, Kinlochleven (1904–1909) (© Alex Gillespie)

Lining a shaft at Kinlochleven, with workers' accommodation in background (1904–1909) (© Alex Gillespie)

Construction of pipes delivering water to Kinlochleven Power Station (1904–1909) (© Alex Gillespie)

Turbine base at intake works, Mullandoch Power Station

Mullardoch Dam under construction, looking downstream, 1950

Drilling prior to grouting on south wing of Mullardoch Dam, March 1951

Plate-laying (W Ralston)

Face-drilling at intake works, Fasnakyle Power Station

Buttressing work on Mullardoch Dam, July 1951

Air pipe and pit props at mouth of main Glenmoriston tunnel

Looking downstream from main tunnel, Invergarry Dam

Loco rails, air pipe and switchboard, main driveway tunnel

Tunnellers' tea break (© Aberdeen Journals)

Muck-shifting: full skips waiting to be removed by locos

Glenmoriston Power Station with pneumatic air line on left; start of shuttering work in background

Drilling under protective steel ring with banking bars and steel packing

Miners waiting to change shift at Mullardoch Tunnel sluice gate

Reinforcement grouting in progress, Mullardoch Power Station

James Williamson cutting the tape at the Sloy Tunnel breakthrough, 12 April 1949
(reproduced by kind permission of Mott MacDonald Ltd)

Tom Johnston inaugurating work on nosheb construction scheme

'World tunnelling record, 557 ft in seven days, 8' 6" bedwidth tunnel, 20–27 October 1955' St Fillans, Perthshire; Bob Sim is 6th from left, first row sitting (Andy Anderson)

Tom Johnston with the locals, turning on the power

Using burning gear on steel

Site clearance (W Ralston)

Clearing a sump (© Alex C Cowper)

Cannich camp

Half of spiral casing for turbine in transit to power station

Mullardoch Dam overtopping, December 1954

Cuileig Hydro Scheme, Construction of Cut and Cover Power Station

Kingairloch Hydro Scheme, Borrow Pit and existing reservoir

Helen Fullarton

I am indebted to Helen Fullarton on two very important counts. First, it's good to get a female face in the hydro picture. Helen, too, is a great adventurer. A science student at Glasgow University in the early 1950s, she took a vacation job working for the caterers at the Shira Dam contractor's camp. After a disagreement with her employer, she returned to Glasgow, bought a van and set herself up as a mobile shopkeeper. She went back to the Shira camp and set about undercutting the camp contractor on cigarettes and other supplies that the men would otherwise have to travel an impossible distance to buy. The contractor had a skinflint reputation for providing inadequate accommodation and driving a hard bargain. With the dam site nearly 3,000 feet above sea level in an area chosen for its annual 105 inches of rainfall, life was hard for the men. Helen's great contribution to our narrative – and my second debt to her – is her eye-witness account of the labourers' experience on the NoSHEB Development Plan.

Helen's life has always been interesting, to say the least. Musician, poet, writer, farmer and environmental activist, she wrote a song called 'The Shira Dam' which draws a truly vivid picture of all the privations capital will inflict on labour if the workers have no union to look after their interests. 'The Shira Dam' describes the absolute powers of the contractors during the NoSHEB Development Plan. If a contractor were determined to keep costs to a minimum, the men would soon know about it, via their digestive tracts at least.

The song is based on what Helen saw at the Shira camp. She met and worked with travelling people at Shira and wrote another of her powerful songs about love between a traveller girl and an Irish labourer. Her song about the Shira Dam is an eloquent reminder of what daily life was like for the men who were working on the Shira Dam. There are few first-hand records of the NoSHEB Development Plan from labour's point of view. As already noted, the men who died in the course of its execution are virtually forgotten now.

Since the beginnings of large-scale hydro-power development, mechanisation, especially of earth-moving tasks, had led to a great reduction in the hard physical effort demanded of labourers in hydro power development. Camp conditions had also improved dramatically since Patrick MacGill's days at Kinlochleven. However, increasingly sophisticated mechanisation called for the men's intelligent and responsible application. The job was still dangerous, above and below ground, always took place in remote locations, and certain contractors did not consider the workers' safety to be their most important operational concern.

Helen Fullarton is one of many who remember the hydro boys with admiration for their courage and strength and with great compassion for their vulnerability. 'The Shira Dam' is a rare and authentic record of their experiences at the hands of an unscrupulous contractor.

THE SHIRA DAM
(Helen Fullarton)

There's a place that's overgrown wi' green at the foot of the Shira Glen
Eleven years a home from home for Carmichael's men.
We came in tens o' thousands to build the Shira Dam,
And the gaithering o' a fortune was every navvy's plan.

I workit in the tunnel and I workit in the shaft,
And then I poured the main dam, it was there I did me graft.
The nipper makes a fortune, a-stewin' up yer tea,
I think he boils his underwear, for it tastes like that to me.

If the gaffer disnae like yer face then it's 'Paddy, are ye tired?
I'll keep ye frae the roarin' rain, get doon the hill, ye're fired!'
But if yer face it's made tae fit, ye'll work the winter through,
And what ye make in the wind and rain, ye'll melt in the mountain dew.

And when ye're doon the glen again ye join a dinner queue,
And at the end a grisly lump – I heard them ca' it stew,
McKay's fat dog it gets the meat, and the milk it's watered sair,
And the soup comes up in the same old pail that's went tae wash the
flair.

The Shira hasnae a Union though I mind when it was tried;
Carmichael came to the meetin' and got on a chair and cried:
'There's no barbed wire around this place, so get ye up the hill,
If you don't like it, jack up boys, your places I can fill.'

But that day we had chicken and the next day we had meat;
The third they took our spokesmen and kicked them on the street.
Aye, on a simmer's evening we built the Shira Dam,
And if they ask you what we used just tell them spam and jam.

The swan it cries on Lochan Dubh and the seagull on the sea,
And city lights and clachan lights are burning merrily.
The Shira Dam's a bonny dam and nothing more remains,
Of the lads who died a-buildin' her but I could gie ye a' their names.

Glenstrathfarrar

If your home is bombed, burnt or blown over then it can be replaced. But with a dam, all hope is gone.
Iain Mackay

By the middle of the 1950s, the NOSHEB Development Plan was being implemented all over the Highlands. The Board was determined to take its exploitation of northern water power resources as far as it possibly could. Plans existed for plenty more schemes than those already completed; James Williamson had outlined 102 in his report to the Cooper Committee. The hydro fanatics on the Board naturally wanted to maximise the output of the NOSHEB system.

Plans existed for a second stage in the hydro-electric development of the River Beauly and its tributary, the Farrar. The proposals would add four generating stations to the existing Affric-Beauly scheme. However, the NOSHEB's plans were to meet with the same determined opposition that had paralysed hydro development twenty years before. After the Board's victory at the Pitlochry Inquiry, its activities had been allowed to proceed more or less unchallenged. But when the Board published the outline of its Strathfarrar and Kilmorack Project (Constructional Scheme No. 30), the Conservative Secretary of State, John Maclay received several official objections. He therefore ordered a Public Inquiry to determine if the Board should be allowed to proceed. Most of the objections at the Inquiry were concerned with the plans for flooding Loch Monar and other potential damage to the environment which Board plans threatened to inflict on Strathfarrar.

Loch Monar lies at the head of Glenstrathfarrar, one of Scotland's most distinctively attractive glens. The River Farrar is a major tributary of the River Beauly, which it joins at the site of thirteenth century Erchless Castle. Glenstrathfarrar is nearly twenty miles long and, as its singular name suggests, contains significant areas of both low-lying arable land and hill pastures. Strathfarrar also has several remnant stands of the Caledonian Pine Forest which once covered much of the Highlands.

In the 1950s, Strathfarrar was used mainly for sheep farming and also for deer-stalking and salmon-fishing. Three proprietors controlled Strathfarrar. Sir Robert Spencer-Nairn, whose family also had property in Fife, owned the eastern end of the glen. Moving west up the Farrar, the Lovat Frasers of Beaufort owned the next section of Strathfarrar. The western end of the glen belonged to Sir John Stirling who used his Strathfarrar property as an extension of his estate at Fairburn near Muir of Ord, which was some twenty miles to the north-west. There had been sheep at Strathmore since 1800. Iain Thomson, in his memoir, *Isolation Shepherd*, has lovingly recorded the remote wild beauty of Loch Monar and the way life was lived in the tiny settlements on its shores. Thomson lived with his family at Strathmore on the north side of the loch and looked after the sheep and cattle driven there from Fairburn. *Isolation Shepherd* paints a poignantly unforgettable picture of the challenges and rewards dictated by an existence surrounded by the intense natural beauty of landscape and wildlife. Thomson's narrative shows very clearly that, despite the community's remoteness, the quality of life enjoyed by its members was very rich indeed.

In 1957, Loch Monar lay at the end of nearly twenty miles of rough track which in some places was no more than bare rock. None of the houses there had a telephone and the post only came to the road-end three times a week. But the loch and its environs supplied the human population with an abundance of natural resources. Iain Mackay, who lived across the loch at Pait with his family told me all about it: 'Even in war-time, we lacked for nothing with our own venison, fish, eggs, vegetables and milk for cheese and butter'. Iain, his brother Kenny and their father supplemented the croft's income by working for Fairburn as stalkers, gillies and stockmen when the Stirlings needed extra manpower.

The Inquiry was held in Edinburgh in the autumn of 1957. Sir Robert Spencer-Nairn was the only landowner to object officially, although both Lovat and Stirling were known to have serious anxieties about the damage that the Board's activities might do to the glen. During the early stages of the NOSHEB Development Plan, Lovat had complained loudly in the Scottish press about the flooding of Highland glens to make cheap hydro-electricity for 'factories in the south'. But by the time the Board announced its plans for Strathfarrar, both landowners were actively involved in local government and so could not openly oppose a measure which promised so much for Highland employment prospects.

However, Lord Lovat's original distaste for hydro development did not stop him accepting a huge cash payment in recompense for damage done by the Board's contractors to nearly twenty miles of salmon river and 650 acres of agricultural land. Malcolm Macmillan, Labour MP for the Western Isles, claimed in a heated Parliamentary debate that Lovat had been 'bought off' by the Board's 'extravagant' compensation award. The press decided to follow the dispute and featured self-righteously indignant denials by Lovat in its pages. His claims that 'I Was Not Bought Off' headlined big photographs of his reassuringly patrician features. He branded Macmillan 'a mud-slinging socialist' and denied he had received as much as £200,000 compensation from the Board. Board officials supported this denial but the subsequent revelation that they had paid Lovat £100,000 only added fat to the fire. As far as Macmillan was concerned, even this amount was too much when the Hydro was blaming 'financial difficulties' for its delay in connecting Barra and other Hebridean islands to the Grid.

Sir Robert Spencer-Nairn was the only Strathfarrar laird to voice his opposition to the Board's plans at the Inquiry. Other official objectors included the distinguished academics, T Elder Dickson, Vice-President of Edinburgh College of Art and Dr WL Edge, Reader in Mathematics at Edinburgh University. Their declared aim at the Inquiry was to save 'the last of the great unspoiled glens'. Tom Weir, representing the mountaineers, insisted that the Board's plans would desecrate 'the monument' of Glenstrathfarrar.

The objectors re-stated all the traditional arguments against the hydro-electric development of the Highland glens. They questioned the ethics of disrupting the life of an existing community for the ultimate benefit of distant electricity consumers. They also claimed that over-capacity within the electricity generating system made the further development of the Affric-Beauly catchment area unnecessary.

Iain Mackay spoke up for his family and Monar neighbours at the Inquiry. They had all signed a petition protesting against the flooding of the Loch. Iain told the Inquiry how the flooding would wreck the homes and the livelihoods of all the people at the west end of Strathfarrar. 'I tried to make them see that for us it was just like the Clearances all over again. We were being treated like pests.'

Tom Johnston was the Board's star witness at the Inquiry. He stressed the valuable way hydro power could contribute to the British economy by reducing the national coal bill. He also repeated the Board's promise that hydro power development would bring 'Joy to the jobless in the glens' as one newspaper article proclaimed during its extensive coverage of the Inquiry. The chairman expressed his sympathy for the Mackay family but regretted that, in his opinion, 'the public interest which requires the Scheme outweighs the purely private interest of Mr Mackay'.

The law took a similarly dismissive approach to the fate of the Pait tenants. Even without confusing the calculations with comparisons to the Lovat haul, it is difficult to imagine a fair figure for the loss to the Mackays of their house, their jobs and, as Iain put it to me, 'our whole world'.

When the Board had secured a favourable outcome at the Inquiry, Sir John Stirling felt able to admit his real feelings to the press about the development of Constructional Scheme Number 30: 'I cannot help regretting it. We did not oppose the scheme because it really is in the national interest and it would be wrong of individuals to try and stop it. I cannot say that I am at all pleased at this ruination of my property. But I have to accept the views of the powers that be that it is a good thing to have all this power developed.' Sir John's remorse must have been made easier by the compensation he received from the Board for the flooding of his Monar properties. The amount was never made public.

There was an inevitability about the Board's victory. It had used its vastly superior resources and organisation to orchestrate a highly effective propaganda campaign throughout the Highlands. As Iain Mackay remembers ruefully, promises of electricity costing next to nothing and the prospect of increased employment opportunities won over the townspeople of busier places like Dingwall and Inverness. 'They swallowed the Hydro Board's propaganda, hook, line and sinker. They didn't really care about what was going on out in the hills.'

Even the Society for the Protection of Rural Scotland, which had fought so effectively to stop the Grampian Electric Supply company from developing Glen Affric before 1943, allowed themselves to be turned back from a fact-finding mission to Loch Monar. Perhaps it was the prospect of a long and uncomfortable trip which helped to dissuade them.

Many people felt that what happened at Pait was symbolic of the death blow dealt by the NOSHEB to what remained of traditional Highland life. From 1943 onwards, the mere possibility of hydro-electric development had exerted a depressing effect on the activities of lairds and crofters alike in glens which were known to be under consideration for hydro power development. As one crofter remembers, 'What was the point of spending money on maintenance or improvement or making any attempt to reverse years of war-time neglect if the whole place might finish up underwater?'

Hydro power development destroyed some of the finest arable land in the Highland glens. Loch Monar was typical of many large Highland lochs in being surrounded by a sizeable margin of fertile land. It wasn't just the loss of this tangible asset that so depressed communities throughout the Highlands. The creation of the NOSHEB reservoirs meant the flooding of over 60 dwelling places and serious damage was done to ancient communication routes such as old tracks and drove roads. A few were relocated but many were not. The network by which folk had navigated the Highlands for centuries was damaged and the viability of many remote places was undermined.

Ironically, the planning regulations which bedevil today's hydro developers would never have allowed the flooding of the Monar marshland with its islands and streams. Loch Monar was one of the most important places in Britain for moor and water-fowl. Teal, widgeon, redshank, greenshank, curlew, snow bunting and lapwings were only some of the regular visitors to the loch before the creation of the reservoir destroyed their habitat.

This sterilisation of the Highland heartlands may well, of course have been inevitable with or without hydro power development. The self-sufficiency of the Monar community might not, for all its riches, have been able to survive the many other changes to affect Highland life from the 1950s onwards. Modernisation was an essential prerequisite to economic development. Highlanders were right to support the promise of jobs and development, for without the NOSHEB's subsidised connection programme the region might have missed out completely on the benefits which progress has bestowed on the rest of Scotland.

But today, the remotely controlled generation of hydro power gives employment to nobody in Strathfarrar. And anyone who has ever felt a vital link with a place they call home could not fail to sympathise with Kenny Mackay senior, Iain's father. He and his wife left the croft at Pait, their home for over 40 years, before the contractor's men began preparations for the creation of the reservoir. The smaller houses were blown up to prevent parts of them disintegrating underwater and then blocking the turbines. The more substantial buildings were simply flooded. Pait Lodge and the Monar keeper's house survived to look out over the new reservoir. But all other signs of the lochside community, which had been home to ten or more families at one time, were lost for ever under the rising waters.

At the new Mackay home further down Strathglass, old Kenny gave Iain Thomson his verdict on Progress: 'Mind you boy, the light in the byre is fine but all the power in the world couldn't replace the pleasure of a fine day on Loch Monar'.

CHAPTER 6

Mackenzie: the Beginning of the End

In 20, 30 and 50 years hence people... will say how tremendously fortunate it was that the water power development took place when it did.

Sir Christopher Hinton, opening the Glenmoriston Scheme in 1958 (quoted in *The Hydro* by Peter Payne)

The NOSHEB's victory in Glenstrathfarrar marked the high point of its political influence. The Board had its own way with the development of Loch Monar in Constructional Scheme Number 30, Strathfarrar and Kilmorack Project. However, this would be the Board's last conventional hydro project to be sanctioned by central government. (Conventional hydro means hydro which does not use pumped storage.) By 1960, out of the 102 potential hydro schemes identified by James Williamson for the Cooper Committee, there were still over 50 left undeveloped. But times were changing and they were changing in ways that would fatally weaken the NOSHEB's political position. New technology and new politics made new enemies for the Board and many of the old ones had never gone away.

At the start of the 1960s, nationalisation was no longer the brand new instrument of socialist re-construction that had changed the face of British industry after the war. State sponsored capitalism was now a fact of political and economic life. The Conservative Party and its supporters considered it their political duty constantly to scrutinise and challenge the operations of the nationalised industries set up by their Labour opponents. The NOSHEB and its Social Clause attracted particularly fierce criticism.

By 1955, the whole of Scotland's electricity supply industry had been nationalised. In 1954, the Electricity Reorganisation (Scotland) Act had given the NOSHEB a new neighbour, the South of Scotland Electricity Board. The SSEB was responsible for generation and transmission of electricity in southern Scotland and the 1954 Act obliged the SSEB to purchase a set amount of electricity from the NOSHEB. This arrangement suited the NOSHEB well, giving it a guaranteed market for easily produced peak load units. But within a year the commercial relationship between the two Boards was put under severe strain, not by politics or economics but by the weather.

In 1955, the Highlands suffered a severe drought. Every single NOSHEB rain gauge recorded a marked reduction in rainfall; some areas were down by one quarter of expected volume. That year, Rannoch Power Station had the lowest rainfall since the Grampian Electricity Supply Company started records in 1931. The drought damaged the working relationship between the North and South Boards as the NOSHEB was forced to rely on its thermal stations at Dundee and Carolina Port to make up the shortfall in its production caused by the lack of rain. Coal-fired power stations produced peak-time electricity more expensively than hydro powered ones and so the SSEB was forced to pay more for the supplies of current it was legally obliged to purchase from the NOSHEB.

The drought was over by 1956 and the working relationship between the two Boards was not seriously affected in the short term. However, seeds of doubt had been sown in the minds of the SSEB managers. They began to question the wisdom of relying on the NOSHEB for a supply of current that might prove unreliable and expensive. The NOSHEB could now no longer assume that it had the unconditional support of the SSEB despite the two Boards' shared status as nationalised utility companies.

In 1957, the Conservative government set up a Parliamentary Select Committee to examine the performance of nationalised industries. During their deliberations, members of the Committee read a paper by Aberdeen University economist Denys Munby that challenged the very fundamentals of the NOSHEB project. Munby, sensibly enough, wanted the adoption of an integrated energy policy that would treat questions of power generation and its costs on the basis of national and long-term considerations. But he could see no place for hydro power in such a policy, believing that it was more expensive, in real terms, than power produced by thermal stations. He compared the costs of different power generating systems and his conclusions did not favour hydro.

The study of economics is not always as scientific as it claims to be; its conclusions can be based on subjective and biased assumptions. For his cost analysis of hydro-electric power generation, Munby made certain assumptions about amortization rates and the real price of base load units of electricity (generated at times of minimum demand) and peak load units (generated at times of maximum demand). Amortization is the writing off of the initial cost of an asset, like a dam or power station, by instalments paid over time. These assumptions were central to the calculations which Munby used to prove the relative costliness of hydro power. Yet these assumptions were not impartial.

At the end of the 1950s, nuclear power was the great technical hope of the future, promising to generate electricity which would be too cheap to meter. Munby's paper claimed that nuclear power would make hydro-electric stations 'built at a very great capital cost... as obsolete as the horse and carriage in fifty years time'. Munby, like most people at the time, was ignorant of the huge cash and environmental costs that nuclear power generation would ultimately involve.

Munby also criticised the NOSHEB for what he considered its unfair pricing policy. The Board charged a uniform tariff for the electricity it supplied to all its customers in the Highlands. This meant that the NOSHEB consumers who lived in remote places on the north and west coasts and in the islands were, in effect, being subsidised by those people who lived in the far more accessible coastal belt of the eastern Highlands. This *de facto* price subsidy was the most effective legacy of Johnston's Social Clause; no wonder that Munby didn't like it.

The fact that this important committee had picked Munby's paper out of academic obscurity was bad news for the NOSHEB. All hydro power's opponents, new and old alike, now had expert opinion to use as ammunition in their campaigns against future hydro development in the Highlands. Among these opponents at the time, the coal lobby was the most vigorous and well organised. The NOSHEB had always boasted about its capacity to reduce the national coal bill. In the austerity of 1947, with its icy winter, fuel crisis and miners' strike, such a boast made excellent publicity. But fifteen years later, with the hardships of the post-war period almost forgotten, the coal industry opposed any method of power generation which reduced the importance of coal to the national economy.

During the 1950s, the coal industry was vociferously defended in Parliament by the Tory MP, Gerald Nabarro. The epitome of the independent backbencher, Nabarro always took a keen interest in the fuel and power industries and was an outspoken critic of all the nationalised utility industries. In 1952, he had set out his views on the need for a national fuel policy in a pamphlet entitled, *Ten Steps to Power*. In it he warned that future hydro-electric schemes 'saving as they do only a tiny amount of coal in relation to the vast capital investment envisaged must be treated with the strictest reserve'. Five years later, Munby's dissertation on the relative costs of hydro power gave Nabarro and the coal industry's case expert credibility. Munby's assertions about the real costs of hydro power seriously undermined the NOSHEB's claims to be making a key contribution to the economy by cutting the national coal bill. Munby's paper also gave encouragement to a right-wing group called Aims of Industry. One of this group's chief goals was to discredit nationalised industries. The NOSHEB, with its socially-inspired pricing structure and its Social Clause, inspired particularly fierce criticism from Aims of Industry and its Conservative Party allies. Moreover, opponents of hydro power now had a crucially important ally: the Treasury itself was growing increasingly unhappy about the mounting expenditure required by the Board's Development Plan and its continuing connections programme. Such concern was only intensified by Munby's claims.

So by 1960, this groundswell of opposition to the NOSHEB's activities had gathered considerable strength. Even in the Highlands, the Board had some critics. Although most people in the north were keen supporters of the Board's activities, loud complaints had been made there that the programme of connections to the Grid was not being carried out quickly enough. Barra, the constituency of Malcolm Macmillan, Lord Lovat's accuser, was only one of the islands where people's patience was being strained by the long wait for electricity. But despite these criticisms, the Board continued to work for the fulfillment of its original aims.

Angus Fulton, the Board's Chief Executive from 1955 to 1966 and a 'hydro fanatic', was determined to exploit every drop of Highland water to the full. In 1960, the Board published its Constructional Scheme Number 37. This scheme was going to use the waters of the River Nevis to operate an underground power station on Loch Leven. The current produced would initially contribute to the public supply and later be made available to industrial users in Fort William.

Scheme Number 37 met with numerous objections, including one from the National Trust for Scotland. But in 1961, instead of ordering the usual Public Inquiry, the Scottish Secretary of State, John Maclay, under direction from the Conservative Cabinet's Economic Policy Committee made a momentous decision. He announced that before a verdict could be given on the Nevis Scheme, wider questions of policy regarding the whole of the Scottish electricity supply industry should be reviewed by Parliamentary Committee.

The Conservative government's Economic Policy Committee chose as chairman of the committee undertaking the review, Colin Mackenzie, a former chairman of the Scottish Federation of British Industries. Serving under him were EW Craig, an Aberdeenshire trades unionist, James Craig, County Clerk of Aberdeen, Sir Josiah Eccles who brought to the committee's deliberations a practical insight born of a lifetime's career in the power generation industry, JS Grant, editor of the *Stornoway Gazette*, JA Keyden, a senior executive from the Scottish steel industry and Professor Alan Peacock, an Edinburgh University economist.

The Mackenzie Committee's brief was officially outlined thus: 'To review the arrangements for generating and distributing electricity in Scotland having regard to (i) the availability and cost of hydro-electric power and other sources of electricity, (ii) the rate of increase in the demand for electricity and (iii) the needs of the remote areas and to make recommendations'.

In effect, the future of the North of Scotland Hydro-Electric Board was now in the hands of government officials and the advisers chosen for them by the Conservative Cabinet. Tom Johnston had retired and the Board had lost much of its influential political support. To make matters worse, it was not able to engage the services of its usual QC, who was well-versed in the technical and economic complexities of the Board's business.

The Mackenzie Committee had to resolve certain central issues in order to achieve its aims. The two Scottish Electricity Boards used five sources of energy to power their generation of current: oil, water, gas, coal and nuclear fission. The committee had, therefore, to determine the relative costs involved in using these different means of power generation and on the basis of these relative costs it had to decide on the size of the contribution each source of energy should make to Scotland's overall generation of electricity. The committee was dependent on the NOSHEB and the SSEB for much of the highly detailed technical and economic data required for making such comparisons.

Inevitably, the information supplied by each Board reflected its own individual agenda. The NOSHEB wanted to maintain its operational and commercial independence and continue building its Development Plan. The SSEB was irritated by its statutory obligations to purchase what it felt was over-priced current from the NOSHEB, preferring to avoid such outlay by building new thermal stations of its own. To many SSEB officials, a merger of the two Boards seemed the most efficient arrangement and one that would give the SSEB the upper hand, thanks to that Board's greater area and customer-population. Amalgamation was also favoured by some at the Scottish Office where civil servants were tired of having to adjudicate between the two Boards and their requests for government funding.

However, public opinion throughout the Highland population was dismayed by the prospect of the NOSHEB's demise. At this time most Highlanders saw the Board as an organisation which was on their side, politically and economically. The Glasgow MP, Bruce Millan, had explained this loyalty to Parliament in November 1962: 'People in the north of Scotland have had very little about which to be optimistic or hopeful over the last ten years, and one of the few things that has given them any grounds for optimism has been the existence of the Hydro-Electric Board and the work it has done in the north of Scotland.' All the major Scottish newspapers vigorously opposed the extinction of the Board apart from those owned by DC Thomson in Dundee. Hamish Mackinven, then Assistant Information Officer for the NOSHEB, wrote to GM

Thomson, MP for Dundee East, pointing out that since the ordering of the Inquiry, 'The Board has never had more friends'. Board members, including Angus Fulton, Ken Vernon and Peter Aitken, worked hard behind the scenes to capitalise on this popular support.

The stringent comparison of the efficiencies achieved by the four different fuels involved masses of extremely complex technical and economic information, some of which may not always have been presented by the two Boards in a strictly impartial fashion. One of the most crucial points of the debate was the nature of the commercial relationship between the NOSHEB and the SSEB. How much better off would the southern Board be if it was allowed to build its own thermal power stations, thus doing away with the need to buy the NOSHEB's surplus current? And how would the Scottish consumer fare if the Boards did amalgamate?

The Mackenzie Committee found it very difficult to reconcile the figures and arguments put forward by each Board and this very difficulty pushed its members towards the idea of amalgamation. Its final report regretted its inability to achieve co-operation between the two Boards and also complained 'how difficult it can be, under the present system, to reach conclusions which are agreed to by both Boards and are valid for Scotland as a whole'.

Eventually, after eighteen months of hearing evidence from both Boards as well as from consumers, local government and industry, the Mackenzie Committee reached a conclusion. It decided that the Scottish consumer would ultimately be disadvantaged if the two Boards continued to be managed according to their own individual interests, rather than for the good of the country. Once this decision was taken, the committee's conclusion was inevitable: 'There shall be established a single authority for the whole of Scotland to be known as the Scottish Electricity Board, consisting of persons appointed by the Secretary of State, that the two existing Boards shall be dissolved; and that all property, rights and obligations of the North of Scotland Hydro-Electric Board and of the South of Scotland Electricity Board shall be transferred to the Scottish Electricity Board.' The setting up of the Mackenzie Committee had prompted widespread expressions of support for 'the Hydro' as the NOSHEB was known throughout the Highlands. In August 1962, the Committee's report calling for the NOSHEB's extinction was published and it triggered even more vehement reactions in the north. Bitter complaints were made in the Highlands where the Boards' disappearance threatened hopes for future industrial development. As one ex-Board member wrote at the time: 'I am afraid if they carried [the amalgamation] out, it would be the death knell of our Highland ambitions.' (quoted in *The Hydro* by Peter Payne). The Scottish media and numerous MPs expressed anxiety about the lack of any other agency capable of continuing Johnston's campaign to modernise the Highlands.

Hamish Mackinven was not solely responsible for the uproar launched in defence of the Board once Mackenzie's recommendations became known. The outcry was too mighty to have been the work of one man and the heroic efforts made by Board members must not be forgotten. But Mackinven was an experienced journalist who had worked as Press Officer for the British Labour Party in the early 1950s. His extensive political and professional contacts made him the right person in exactly the right place in the summer of 1962. From the start of Mackenzie's deliberations he had been 'feeding stuff left, right and centre' in an intense effort to promote the Board's cause. His propaganda campaign intensified with Mackenzie's decision in favour of a single Scottish Electricity Board. Mackinven recalled his desperation: 'The South Board was sure they had got in the bag and I thought: 'What can I do now?'.'

With a contact in the Scottish National Party, Mackinven helped to stage a *de facto* Highland referendum on the fate of the NOSHEB. Hamish's contact wore two official hats: he was a deputy secretary at the Board and also secretary of the Electricity Consultative Council (North of Scotland). He was able to convene a meeting of the Electricity Consultative Council to voice its official opinion on Mackenzie's proposed abolition of the NOSHEB. The members of the ECC came from all over the Highlands and Islands and included people from all social classes. It was considered to be in genuinely close touch with public opinion in the north. The Council called a meeting with only one item on the agenda: the need to rally support for the Hydro Board in its fight for survival.

The Electricity Consultative Council produced a report on Mackenzie's findings which achieved this purpose perfectly. The Council made a wide range of points in the Board's defence: the

Board had arrested economic decline in the Highlands; it had effectively managed both the large and small scale elements of its Highland operations; it was not in debt; and, it had achieved over 90% of its proposed connections to the Grid. Moreover, as the Council's report went on, supplies from the Scottish coalfields were 'problematic' and it maintained that 'the north will be best served by the maximum development of hydro power in the shortest possible time.' Like many others then and since, the Council questioned the assumptions made by the Mackenzie Committee about key economic variables including forecasts of bank interest rates and the price and availability of coal. Above all, the Consultative Council stressed that the NOSHEB had the full backing of public opinion in the north. The ECC report declared that the Board had succeeded as a 'Highland enterprise with the interests of the Highlands and Islands as its primary objective'.

This apparently heartfelt unanimity of Highland support and affection for the Board may have surprised the Conservative Government, all those miles away in London. I think it is safe to assume that the Cabinet Committee on Economic Policy which had set up the Mackenzie Committee and had, to some extent, influenced its conclusions, had not expected the NOSHEB to survive the committee process. In any event, the Electricity Consultative Council's unambiguous support of the Board was made in a political vacuum with the SSEB management divided over the advisability of a single authority and the Scottish Office was also split at the most senior level on the amalgamation issue. The government had agreed to contribute significant resources to speed up the connection programme, very much to console the north for the loss of its Board. The Treasury, therefore, was reluctant to see the Scottish *status quo* disturbed. Most important of all, a General Election was imminent and the people of the Highlands had made their wishes known. So, in July 1963, Michael Noble, who had succeeded John Maclay as Scottish Secretary of State the previous year, announced in Parliament that the amalgamation of the North and South Boards proposed by Mackenzie would not now take place.

There was great rejoicing throughout the Highlands at this reversal of the Mackenzie Committee's preference for a single

Scottish Electricity Board. Senior NOSHEB officials, however, were well aware that the rest of the committee's recommendations had not been overturned. Most important of these was Mackenzie's insistence that any hydro plants built in the future would only secure governmental approval if they could guarantee a satisfactory economic performance. This performance would, henceforth, be judged in strict comparison with other available forms of power generation. If the Government was going to invest in any form of power generation, it wanted to be sure that it was spending money in the wisest possible way.

Mackenzie had at least accepted that limited scope did exist for further conventional hydro development in the north. The Nevis scheme was pending and in April 1962, the Board published details of a Constructional Scheme at Laidon on Rannoch Moor. In March of the following year, Constructional Schemes Nos. 33 and 39 were announced for Fada-Fionn and Loch a' Bhraoin in Wester Ross.

The Board was confident that its plans would meet the economic performance criteria laid down by Mackenzie and pressed the Scottish Office to set up a Public Inquiry into Fada-Fionn and Laidon. The Board had to wait until December 1963 for the Secretary of State, Michael Noble, to order an Inquiry into the largest of the proposed schemes, Fada-Fionn. The Inquiry sat from January until March 1964 and the report it published in November 1965 ran to over 2,000 pages. The Reporters were JA Dick QC and Professor AD Campbell, economist, of St Andrews University. A young civil engineer who attended the Inquiry still carries a clear mental picture of the two Reporters setting off on donkeys for a first-hand look at the site of the proposed scheme. They made a strange couple as he recalled for me: one Reporter was rather a small man and the other was huge.

There were no objections lodged at the Inquiry by public bodies; even the Association for the Preservation of Rural Scotland and the mountaineering clubs made no official complaint. The ten official objectors represented private fishing and stalking interests. One of these, Col. William H Whitbread, was a prominent member of Aims of Industry. He brought two academic economists to the Inquiry to argue that hydro development at Fada-Fionn was not in the national economic interest. One of them, Colin Clark of Oxford University, went so far as to claim that none of the NoSHEB hydro plants had ever made a profit. The economists once more put the blame on hydro's high capital costs.

The Board put up a good fight, insisting that none of the ten objectors were genuinely concerned with the national interest. Arguing for the development at Fada-Fionn, the Board also stressed the practical difficulties involved in the opposition's alternative proposals for supplying the Highlands with the power which Fada-Fionn would have generated. Such a supply could only be provided by importing current from large thermal stations in the south of Scotland. Transmission over such a huge area and over such difficult terrain was bound, the Board argued, to be problematic and would inevitably involve the extensive use of overhead cables on pylons, an increasingly unpopular sight at the start of the 1960s, especially among the wilderness lobby. The Board also pointed to the benefits that would accrue across the region if its expansion programme were allowed to proceed. Revenue-earning opportunities must not be missed if the Board's connection programme were to be adequately financed.

When the results of the Inquiry were finally announced in November 1965, the Board's hopes of continuing with its limited plans for building more conventional hydro schemes were completely dashed. The Reporters had decided that Scotland's electricity requirements could be more economically supplied from 'other sources, even allowing for transmission costs'. As Peter Payne would later note sadly, 'It was the end of an era'.

Many ex-NoSHEB employees I spoke to still feel bitter about the way that the Fada-Fionn Inquiry was conducted, in particular the anti-hydro bias behind the way critical economic variables were defined and calculated. Under direct instructions from the Treasury, the Reporters were forced to judge the Schemes' economic performance against a figure of 8% net return on capital expenditure. This arbitrarily high figure was justified by the Reporters' desire to prevent 'the misallocation of resources, [especially when] there are likely to be substantial and increasing demands [in the future] for capital for all sorts of projects in the UK'. I was told more by seriously disappointed NoSHEB men about the 'disgraceful, immoral [way that] fudged figures' were used to kill off the Hydro Board's building programme. As well as the Treasury's insistence on this relatively high net return figure, the Board was faced with the Reporters' absolute refusal to entertain the possibility of significant fuel price rises in competing power generation systems. In less than ten years from the Fada-Fionn decision, coal and oil prices had quadrupled, making nonsense of the claim that hydro could not compete with these systems economically. Energy policy became the political hot potato of the 1970s and fuel crises will continue as long as global energy production relies chiefly on non-sustainable resources. No wonder the Board men were 'pig-sick'.

A month before the report was published, Labour won a General Election victory. Would the NOSHEB's mission be saved at the last minute by the intervention of the new administration? After all, the Board had started life as the first nationalised industry in Britain; it was Labour Party heritage. Surely, a socially motivated public enterprise brought into being by a Labour champion should be preserved?

But the new Prime Minister, Harold Wilson, declined to play Prince Charming to the Board. Famously entranced by 'the white heat of technology', Wilson was captivated, like almost everyone else, by the expansive promises of nuclear power. Ignorance about the ultimate costs of decommissioning nuclear facilities meant that the new government seriously underestimated the relative cheapness of hydro power.

In accordance with Mackenzie's recommendations, the Board was allowed to build two pumped storage hydro schemes. The first was part of the Loch Awe Scheme in Argyll. Cruachan, 'the hollow mountain', is a spectacular world prototype development. The Board also took over BAC's site at Foyers. Their contractors dismantled the existing plant and constructed one with a capacity twenty times larger. The construction teams at both sites had to overcome serious problems and both schemes are still generating a handsome return. Willie Logan, one of the contractors at Foyers, went bust six months into construction and he was killed in 1966 when the private plane flying him home crashed into a hill on the outskirts of Inverness.

CHAPTER 7

Privatisation: Goodbye to the NoSHEB

CRUACHAN AND FOYERS WERE finished more or less on time and, allowing for inflation, within budget. The NoSHEB's thermal and hydro power generating facilities enabled it to contribute, along with the SSEB's power stations, to Scotland's mixed system of thermal, hydro and nuclear power generation.

But the Board also wanted to carry on expanding its use of Highland hydro power resources. Throughout the 1970s, the NoSHEB tried to secure governmental approval for a pumped storage scheme at Craigroyston on the eastern shores of Loch Lomond. In an attempt to win support for the scheme, the Board undertook extensive environmental and economic studies in an attempt to convince politicians and conservationists of the wisdom of its plans. But by 1980, the increase in Scotland's demand for electricity had slowed down markedly and the Board never even submitted a report on Craigroyston to the government.

However, in the previous decade, the power generation industry was affected worldwide by some very important developments. All sorts of assumptions were invalidated by the UK fuel crisis of 1974. People began to realise that the supply of oil from the Middle East would never be completely safe from political manipulation. A proper understanding of the real costs and difficulties inherent in nuclear power generation was also beginning to dawn. There was every reason, then, for the NoSHEB to continue to promote hydro power as a renewable energy source.

But the prevailing political wisdom of the day was not interested in positive, imaginative answers to all-important questions of national and global energy policy. In 1979, Margaret Thatcher became Prime Minister and so began for Britain the agonising process of death by a thousand cuts. In 1983, the NoSHEB tried once more to interest government in hydro development. The Board held extensive local consultations in Wester Ross before submitting a new constructional scheme to the Scottish Office.

The Board had identified six small run-of-river schemes and the first two to be proposed were on the Rivers Grudie and Talladale. The combined cost of these two developments was the relatively small sum of £8.5m and there had been no outstanding objections after the Board satisfied concerns about possible dangers to sea-trout spawning beds. However, George Younger, Secretary of State, refused to approve the scheme because of what he called economic circumstances. I think it might have been more accurate to blame Margaret Thatcher's monetarist priorities for this final dashing of the Hydro Board's hopes.

So the NOSHEB adventure was over; it would never build again thanks to the decimation of government spending. But the Board had endured and had retained its operational independence. Since its inception it had connected most of the Highlands' inhabitants (99.3% by 1989) to the National Grid, overcoming massive logistical and climatic obstacles in the process. In 1984, it started its final Five-Year-Plan to make some of the Highland connections still remaining to be done. The Board had also carried on investigating other ways of generating power. It continued experimenting with wind power and after its plans to commission nuclear power stations in the North-East were rejected by the government, it built an extremely efficient thermal power station fuelled by gas and oil at Peterhead. The Board survived another government inquiry in 1985, conducted this time by the Monopolies and Mergers Commission. The inquiry failed to undermine the Board's dual role in the generation and distribution of electricity. Despite the limits set on its expansion, the Board was in control of a successful enterprise, with a specific product and a well-established demand for that product. All of which, in the Britain of the late 1980s, could mean only one thing: privatisation.

The NOSHEB, Britain's first nationalised industry, had been fighting off right wing attacks ever since its inception. The final, fatal blow was struck as part of the privatisation programme which forcibly replaced the ethos of public service with that of private profit throughout Britain's utilities. The North of Scotland Hydro-Electric Board became Scottish Hydro-Electric plc in 1990.

Highlanders were not altogether impressed with the change. Willie from Lochluichart was now employed by the new company. He could only wonder, when he looked at the photo of Tom

Johnston on the office wall, what on earth the old chairman would have made of this strange successor to his creation. The NOSHEB had been brought down by a government which wanted to denationalise wherever possible. No doubt overmanning and other inefficiencies had been a bad part of the old set-up. However, after the storms, floods and infrastructure worries of the last decade, the evangelical conviction that free market forces would automatically take care of everything now seems more questionable than ever.

So how did privatisation transform a socially motivated enterprise into a profit driven one? What happened to the Social Clause, the emblem of the Board's non-profitmaking commitment to stimulating economic growth in the Highlands? The answer, as is so often the case in the Board's history, hinges on the way the figures were calculated. Apart from the Cruachan Scheme, which was handed over to Scottish Power plc on privatisation, the new plc had exactly the same plant, property, equipment and workforce as the old Board. What then could have happened in the year immediately before privatisation to enable the NOSHEB to record a profit of £50m, with interest charges of £60m paid off? Nothing approaching this sort of capital had been there in the previous year's figures. In fact, this sudden profit was the result of calculating depreciation on the Board's 'civil assets' in a new way. Now, the new plc announced, these assets were 'deemed to have an infinite life since they are maintained'. No depreciation and, hey presto, there's a profit of £50m to encourage potential shareholders.

Privatisation effectively threw out the 1943 Act's social agenda. If the NOSHEB had shown itself to be in possession of such disposable capital whilst operating under the 1943 Act then anyone putting forward a 'Measure for the Economic Development or Social Improvement of the North of Scotland' could have laid claim to some or all of that capital according to the terms of the Social Clause. In the event, precisely the opposite occurred with the new-found cash being used to attract private investment and increase the benefits of privatisation to the Treasury. In 1998 Scottish Hydro-Electric plc merged with Southern Electric and became Scottish and Southern Energy. The new company's generating and distribution activities range across the British Isles from Unst in Shetland to the Isle of Wight.

CHAPTER 8

What Future for Hydro-electricity in Scotland?

PRIVATISATION IS A CENTRAL part of free market economics. Free Traders aim to maximise profits and production for the ultimate benefit of their investors. Free Traders believe that this default growth policy is a valid way of using the earth's natural resources and that there will always be enough profits generated to make good any environmental damage such policy causes. These Free Traders would rather deal with scarcity by increasing prices than by reducing consumption.

The privatised electricity supply industry demonstrates perfectly the radical contradiction between liberal economic objectives and the sustainable use of the earth's resources. In 1926, the Weir Committee encouraged British industry to increase electricity generation and consumption as part of a competitive push for modernisation. At the start of Britain's Electric Age this pioneering state regulation of the supply industry had to demand prodigality. The privatisation of the electricity supply industry continued to sanction and promote free market values in the operation of the UK's utilities. The Conservative government responsible for privatising the electricity supply industry was apparently unconcerned with the problems of pollution and sustainability that such unfettered consumption must inevitably cause. In line with the political dogma which opposed public sector ownership of the infrastructure, the privatised utilities were encouraged to proceed in the headlong pursuit of profits for their new shareholders. The system provided the government with proportionally mighty tax revenues and that was the UK's energy policy at the start of the 1990s. If there was no such thing as Society, as Margaret Thatcher claimed, then Society could hardly be responsible for the Environment, could it?

But if the UK did not have a coherent energy policy designed to protect the environment and achieve sustainability then what was the Non-Fossil Fuel Obligation all about? Part of the 1989 Electricity Act which introduced privatisation, the NFFO certainly sounded Green enough. It appeared to promise support for the use of renewable energies in the generation of the national electricity supply. Renewables did benefit from the NFFO, as we shall see. But there was much more to the Government's NFFO strategy than the support of renewable energy.

This Obligation (Scottish Renewables Obligation in Scotland) required the newly-privatised Regional Electricity Companies to obtain certain amounts of their new generating capacity from non-fossil fuel sources. At that time, electricity generated from renewable energy sources like wind, wave, biomass, waste and hydro, was more expensive than the electricity produced by nuclear power or the combustion of fossil fuels. The NFFO allowed non-fossil fuel generators to be paid premium rates for their products. However, the RECs did not actually have to pay more for this more expensive energy. The cost premium to the RECs of these power purchase agreements was not financed by the government or by the profits of the RECs. The RECs had to pay the baseline cost while the extra cost was met from the Fossil Fuel Levy, a surcharge imposed on domestic and small commercial customers' electricity bills. The amount of the levy was worked out each year to meet these total extra costs of the coming year's NFFO. Each Regional Electricity Company's share of the cost of the Obligation matched its share of the national output. If a company supplied 20% of the national energy total, then it must provide 20% of the Obligation costs. Continuing the doublespeak tradition of the NFFO, the levy was raised on all electricity consumed regardless of how it was generated, rather than the power actually produced from fossil fuels.

The NFFO appeared to be offering the renewable energy sector important support. However, the true purpose of the NFFO was partially obscured by its Green-sounding title. The Tory government's real intention in setting up the NFFO was crucially linked to its privatisation agenda. The NFFO proved that the British ruling establishment was far more interested in protecting commercial interests than in protecting the environment.

The non-fossil fuel which had been the chief inspiration for the NFFO was not in fact any renewable resource but nuclear power.

The accidents at Three Mile Island in 1979 and Chernobyl in 1986 gave the nuclear industry some dreadful publicity. The British government wanted to extend its privatisation of the electricity supply industry to include the nuclear powered portion of that industry. However, the City now had serious misgivings. No new nuclear power stations had been built in the USA since Three Mile Island and the high costs of decommissioning nuclear power plants had been highlighted by the Ukrainian accident. Such realisations endangered the long-term profitability of the nuclear power industry by threatening future investment in it. In Britain the NFFO was devised, first and foremost, to guarantee that very investment. Apparent support for renewable energy gave the government excellent cover for its protection of the nuclear industry. Nuclear power was guaranteed a profitable long-term market thanks to its non-fossil fuel status.

However, despite the pro-nuclear inspiration of the NFFO, renewable energy projects did do well out of the five rounds of price competitions placed under the NFFO from 1990-99 and a further five similar price competitions in Scotland and Northern Ireland (referred to as SRO and NI NFFO). These orders, managed by the Department of Trade and Industry, enabled long-term contracts to be awarded to generating projects fuelled by renewable energy. In Scotland, the rounds of Scottish Renewables Obligation were managed by the Scottish Office. Renewable energy projects all over the UK were able to attract investment based on fifteen year power purchase contracts and agreed prices. Indeed, the shrinking price of electricity generated from renewable energy is increasingly able to compete with the price of power produced from fossil fuels. The projects being built under the NFFO, the SRO and the NI NFFO contracts have been able to take advantage of improvements and developments in these technologies which have resulted in lower generating costs. The NFFO process has indeed helped the commercial development of renewable energy technologies.

But the NFFO legislation offered little direct support to the renewable energy industry beyond the subsidy of its higher prices. Many renewable energy schemes have been refused planning permission and, for much of the decade following privatisation, the state provided scant encouragement to investment in any sector of the renewable energy industry. There was no systematic examination of the full economic potential of renewable energy. Research and development funding was sporadic. The NFFO's method of awarding renewable energy generating contracts in tranches meant that no sustainable market for renewable projects or for equipment used by such projects could emerge. Instead, a stop-start situation prevailed, discouraging all potential investors in hydro and the other renewables. Potential turbine manufacturers, for example, were not attracted to such an unstable market place, being unable to organise a smooth production flow.

Meanwhile, other European countries were funding research and development for their renewable energy industries, in both generating and manufacturing sectors.

Eventually, renewable sources of energy, including hydro power, did ultimately benefit from the non-fossil tariff. These energy sources have become an integral if relatively small part of the privatised power generation scene. Hydro-electric power generates approximately 11% of the electricity generated in Scotland today. Several private hydro schemes have been set up in response to the profit-making opportunities presented by the Scottish Renewables Obligation, Scotland's version of the NFFO.

Private developers like the Dundee-based Highland Light and Power Ltd provide landowners with the technical, planning and investment expertise required to set up small hydro schemes for connection to the Grid. Many Highland landowners are now converts to hydro power, convinced by the profits offered by the SRO. And it's not just the lairds that are cashing in. The Assynt Crofters, who have owned the North Assynt Estate since their celebrated buy-out in 1993, joined forces with Highland Light and Power to develop a community micro-hydro scheme on the Estate – potentially a great example of a hydro power scheme benefiting the very people in whose community it operates. However, the presence on site of a pair of black-throated divers and some freshwater mussels made planning permission very difficult to obtain. Costly modifications had to be made to the original plans in order to satisfy objections from Scottish Natural Heritage; all in all, not a very encouraging precedent for future community-based hydro schemes.

For most of the 1990s, prospects for the UK renewables industry did not exactly sparkle. The political will simply did not exist to change society's free market energy priorities. Would-be investors in the hydro power sector were especially deterred by hydro's high capital cost and uncertainty over planning.

Finally, however, in the decade following electricity privatisation, the threat posed to Western economies by climate change began to be taken seriously by politicians. There was simply too much meteorological evidence to ignore, especially regarding global warming. An excess of man-made carbon dioxide emissions caused by the consumption of fossil fuels was blamed for escalating climate change. Reducing consumption of these fuels was deemed a global priority. In 1997, signatories to the Kyoto Protocol agreed to adopt measures to reduce carbon dioxide emissions in their own countries. Kyoto was immediately criticised for setting low reduction targets and for not making those targets mandatory. But as far as the prospects for the renewable energy industry were concerned, Kyoto inspired a radical change of direction in the UK. When John Prescott negotiated Britain's agreement at Kyoto in 1997, he was ending two decades of monetarist refusal to even contemplate the need for a sustainable energy policy.

Under the Kyoto Treaty, the UK is committed to reducing, by 2010, its emissions of carbon to 20% below its 1990 level. To achieve some of this aim means generating 10% of its electricity from renewable resources by the same deadline. The UK legislation confirming this change of direction is the Utilities Act which received Royal Assent on 28 July 2000. At the Act's cornerstone is the same sort of Obligation as the NFFO which ensures a significant market for renewable energy. Fear of climate change has produced this U-turn as the new terminology indicates. The Non-Fossil Fuel Obligation will be replaced by the Renewables Obligation and the Renewables Obligation (Scotland).

This Obligation on electricity suppliers is different from the NFFO. The NFFO was an obligation to buy green electricity imposed on the electricity companies. The new Renewables Obligations require the electricity suppliers to go and find green electricity in order to be able to achieve a set proportion of their sales output from renewable energy sources. This creates a real business opportunity for the embryonic renewables industry. Electricity supply companies will have to contract with renewable energy companies to create new green electricity generating capacity. There is no levy on customers akin to the Fossil Fuel Levy. There is, however, a new energy tax on non-domestic consumers: the Climate Change Levy.

The CCL was introduced on 1 April 2001. The new Levy operates in the same way as the old NFFO; it is a tax on the consumption of electricity produced by oil, gas and coal. Industrial and commercial customers can reduce their exposure to the new tax by investing in energy efficiency measures to reduce their energy consumption. They can also invest in renewable electricity generation as this green electricity is exempt from the CCL. This favourable investment opportunity will create new demand for renewable generating capacity. Non-domestic customers will avoid paying the levy on electricity if they buy 'green' electricity from an accredited supplier or if they generate their own current from renewable resources. Overall, the Act gives the renewable energy industry a sound future for the first time. The Labour government has already extended the possible lifespan of the new Obligation framework to 2026. There is now good reason to invest in the increase of national renewable energy capacity.

Tom Douglas, Consultant at Mott MacDonald Ltd and formerly a partner in the firm founded by James Williamson, has been biding his time. He was the young civil engineer who watched the Reporters decide to cancel the great NOSHEB building project. His contribution to this book began when he kindly sent me a copy of Donald Hamilton's memoir of James Williamson. He also told me about the 60-odd schemes designed by James Williamson that were not built before the Mackenzie Committee ended the Development Plan. In the Spring of 2001 he sent me his assessment of the Kyoto Treaty effect on the contemporary hydro development industry:

> While there have been numerous reasons in the past for not making use of renewable resources, recent legislation provides a real opportunity to develop a worthwhile, extensive programme of many small to medium sized hydro-electric schemes within the next decade or more.
>
> The principal catalyst is the recent passage of the Utilities Bill. This provides the framework for the cost-effective development of renewable energy schemes in general and hydro in particular. The opportunity will be enhanced by the specific values which are to

be placed on renewable energy units in the supplement expected to be approved by the Scottish Parliament in autumn 2001.

Tom's forecast is already proving correct. Governments that signed the Kyoto Treaty committed their countries to increased use of renewable energy for electricity generation in order to reduce emissions of carbon dioxide and greenhouse gases like sulphur dioxide and nitrous oxide. At the time of writing, the Westminster Government is aiming to generate 20% of the UK's electricity from renewable energy by 2020. The Scottish Executive, on the strength of Scotland's superior renewable energy resources, has set a sub-target for Scotland of 40% by 2020. Such targets will be effective incentives for investment in all renewable energies, including hydro.

This effectiveness is already evident in the Highlands, where there are currently twenty-eight renewable energy schemes seeking planning permission including eight hydro schemes. Scottish and Southern Energy plc has already succeeded in satisfying extremely rigorous environmental planning requirements with two Small Hydro Schemes. One is at Cuileig, on the River Droma, south of Ullapool, and the other is at Kingairloch in Argyll.

In order to obtain planning approval for both schemes, S&SE's designers had to show comprehensive and detailed consideration for all aspects of the existing environment. The legal pressure on developers to minimise the environmental impact of the schemes is reflected in the fact that Scottish Natural Heritage, the Scottish Environmental Protection Agency and the Highland Council were all statutory consultees during the planning and execution of the projects.

Cuileig, a run-of-river scheme, was commissioned in January 2002. S&SE's contractors have kept environmental disruption to a minimum. The main pipeline is underground and the power station is similarly concealed apart from one exposed face that is clad in local stone. The extra cost of situating the power station partly underground is offset by the greater generating efficiency gained due to the increased head. S&SE's scheme at Kingairloch, incorporating dam and reservoir storage, is expected to be in operation by Summer 2004.

But however impressive this swarm of developments and potential developments appears, it is not the result of a sustainable long-term energy policy. Such a policy could pay proper attention to guaranteeing future security of supply while also being committed to developing multiple and flexible responses to the massive problems posed by climate change. Most of these current applications are for Small or Micro Hydro with Installed Capacity of less than 20 MW. The development sites are controlled either by big firms like S&SE or individual proprietors and landowners. All the proposed schemes will capitalise on the premium market for green electricity guaranteed by the Scottish Renewables Obligation. But that encouragement and the renewables targets mark the limit of direct state support for Highland hydro power which is being developed in a piecemeal fashion, without the benefits of systematic and co-ordinated organisation.

In 2003, Scottish and Southern Energy launched one of these Highland applications to build Scotland's first Large Hydro scheme for forty years, at Glendoe, in the western end of the Monadhliath mountains, to the east of Fort Augustus. The firm was understandably keen to emphasise its links with the pioneering past of hydro power in the north. Ian Marchant, Chief Executive, promised that the scheme would be developed 'in keeping with the traditions of our predecessors'. Such assurances make excellent PR, but S&SE, with all its wealth and operational mightiness, is a privately owned company. In contrast, the NOSHEB was founded primarily as a state-sponsored tool for the economic salvation of the Highlands. The NOSHEB tradition rested on the systematically planned and executed exploitation of Highland water power as a key regional and national resource. This resource's strategic importance has become even more valuable with hydro's role as a renewable energy source, generating clean renewable energy. But none of the private firms who have applied for permission to build hydro schemes in the north have ambitions that extend beyond satisfying their shareholders.

Thomas Johnston, Hydro Boy and pragmatist par excellence, would no doubt agree with those who reject the Labour Government's confidence in the power of the free market to safeguard the environment. Tony Blair's foreword to the 2002 Energy Review leaves us in no doubt about his priorities: 'The introduction of liberalised and competitive energy markets in the UK has been a success and this should provide a cornerstone of future policy.' Sorry, Mr Blair, but the environmental crisis will not respond to laissez faire methods. Whatever businesses may say, their primary responsibility is to their shareholders and not to the natural environment or future generations. It is for these future generations' sake that we must have a proper energy policy now to regulate the activities of energy producers and consumers alike.

Many other governments in the liberalised and globalised world economy do not take executive responsibility for their country's electricity generating industry and nor have they faced up realistically to the monumental and problematic issues related to security of supply and climate change. Like Blair, such governments believe that the market will take care of everything and that no restrictions should be imposed that would fetter economic activity and reduce a country's competitiveness. But Adam Smith's Invisible Hand, which he supposed would automatically make market forces work for the common good, was never meant to operate when resources are finite and runaway consumption undesirable. This attachment to market values seems fatally at odds with UK government declarations about dealing comprehensively with the threats posed by climate change and finite fossil fuel reserves. Setting targets for the increased use of renewable energy is as near as the UK Government and the Scottish Executive have come to directing energy consumption. However, many environmentalists believe that the targets set by the Kyoto Treaty are not high enough to mount a worthwhile challenge to climate change. Moreover, these targets are for increasing the proportion of renewable energy used to generate electricity not the proportion of renewables used to generate our entire energy needs. In fact, electricity generation only accounts for 20 – 30% of our energy requirements with the rest divided between fuel for heating and fuel for transport. This means that the 20% target for renewable energy use is in fact nearer 6% of our total energy use. The PR credit that politicians get for setting these targets relies on public confusion about the difference between energy and electricity, confusion that has been allowed to persist in the media despite the urgent importance of thinking through our future energy strategy.

Energy influences every single aspect of a world that thinks itself pretty damn clever until there's a power cut. The Californian power blackouts of 2001 showed exactly what happens when a privatised generating industry fails to make adequate investment in infrastructure. The whole state suffered a series of power cuts. Business was seriously disrupted, apart from in Los Angeles where the electricity generating industry had not been deregulated and where the power supply was maintained.

So for everyone's benefit, and that of Scottish hydro power development, this half-baked reliance on market values should be abandoned. Government must take responsibility for determining and executing energy policy, which should be designed to achieve long-term sustainable security of supply without neglecting the challenges of climate change.

What is needed is the same sort of visionary singlemindedness with which Thomas Johnston achieved the establishment of the NOSHEB. He overcame powerful and long-standing opposition in order 'to do something for the Highlands'. His career was one of devoted public service undertaken against the background of an almost vanished confidence in the power of state planning to get things done and make things better. Johnston's thoroughness and determination gave the NOSHEB nearly twenty years of operational independence, just long enough for the creation of a highly efficient regional power supply system. Johnston's great strength was his practical determination to overcome obstacles and to secure the implementation of effective solutions. His far-sighted yet pragmatic approach is precisely what is needed to replace the current government's refusal to put state authority behind a truly realistic and effective energy policy.

An energy policy designed to replace reliance on the market would start with the creation of an Energy Ministry with enough power and permanence to transcend short-termism born of electoral considerations. Such a body could oversee the design and execution of long-term policies capable of tackling the multifarious problems of establishing a truly sustainable energy policy for the Scotland and the UK.

So, how might such policies work? Starting with the supply side of the energy equation, a proper energy policy could make a vital contribution to ensuring sustainable security of supply. The

2002 Energy Review stresses the dangers to security of supply from dwindling indigenous and other fuel reserves combined with the difficulties of securing imports against the background of heightened international tensions.

UK energy policy could maximise the use of indigenous renewables by planning the systematic and integrated exploitation of such resources. This would be an excellent alternative to the piecemeal, profit-led patterns of investment and development prevalent in the renewables environment today. If hydro development were undertaken by the type of central agency dedicated to increasing use of renewables in Scotland and the UK, then all sorts of benefits would proceed from the systematically organised exploitation of those resources. Operationally, there would be great advantages, including integrated production, design and purchasing. Moreover, a more managed renewables environment could support the creation of an industry to manufacture components for use in renewable energy projects.

Denmark's thoughtful commitment to renewables has had just this effect: Vestas, which started as a blacksmiths and light engineering business, currently has a 23% global market share of wind turbine production. The firm only began wind turbine production in 1979 as a direct response to Denmark's energy policy requirements. In 2002 Vestas opened a wind turbine factory employing over 150 people in Campbeltown on the Kintyre peninsula: a fine example of renewables bringing much-needed employment to rural Scotland. The Scottish Executive has made lots of noise about its intentions to capitalise on this sort of opportunity; the sort of pro-active energy policy I am advocating would make the fulfilment of such aspirations far more likely than at present.

Another supply side context in which our imagined Energy Ministry could play a decisive and very beneficial role is in relation to the costly and time consuming difficulties obtaining planning approval currently experienced by all would-be developers of renewable energy projects, including hydro. The encouragement given to potential developers of renewable energy projects in the 2001 Utilities Bill is not matched by guarantees of a more sympathetic planning regime.

Planning priorities in line with a sustainable energy policy should be established and adhered to. Today's wilderness-loving brigade, regular opponents of hydro power development, should acknowledge the impractical selfishness of their refusal to countenance change: no landscape is exactly as it was in the beginning and choosing conservation of scenery over clean power generation may be a luxury that our society cannot afford.

Obtaining suitable sites for hydro development will always be controversial especially in the Highlands where 'unspoilt scenery' is a vital asset for many businesses. Inevitably, conflicts about land use will waste valuable time and cost money. There must be a more practical way of settling these important and pressing disputes. A proper Scottish energy policy would be committed to making the most of Scotland's renewable energy potential. It could institute a Renewable Energy Commission which would have the job of identifying the best sites for the development of hydro, tidal and wind power development. If the landowner refuses to sell or lease the selected site, the commission, should energy requirements dictate, could obtain it through compulsory purchase. Ownership could then pass on to the local community. The cash dividend might not be huge but at least the locals would derive some benefit from the exploitation of their local resources. There must be a more productive way of dealing with planning for renewable energy projects; an Energy Ministry could establish properly effective priorities for Renewable developments, including hydro. In January 2004, the Scottish Parliament's Enterprise and Culture Committee launched an inquiry into Scotland's renewable energy resources. The inquiry will be canvassing opinion throughout the country and hopes to report by summer 2004. A positive Scottish solution to the planning stalemate would be a marvellous result.

A further major deficiency in government plans for renewable energy development concerns transmission infrastructure. Existing transmission lines in the north of Scotland were originally built to serve a rural area with a relatively low demand for electricity. They were not designed to accommodate the large numbers of requests for connections to the electricity network being made by renewable energy developers today. Scottish and Southern Energy has drawn up plans for new 400,000v electricity transmission lines to export electric current from the Highlands to the central belt of Scotland.

The firm is well aware that such an upgrade is essential if Scotland's ambitions for renewables are to become a reality. The proposed route for the new lines runs through the newly created Cairngorm National Park making the chances of its approval somewhat problematic and highlighting once more the need for a brutally pragmatic energy policy with a clear set of priorities and enough practical determination and resources to maximise the benefits which Scotland and the rest of the UK could derive from development of indigenous renewable energy resources.

A proper energy policy would not just regulate supply. Though the very idea would be anathema to the free traders, a comprehensive energy policy could also intervene by controlling consumption, a very reasonable strategy, given the finite nature of fossil fuel reserves and the pressing need to reduce carbon dioxide and other greenhouse gas emissions. But how could this be achieved?

Our imagined Ministry of Energy would be able to carry out demand side management. This could be undertaken in a number of ways. A fully empowered Ministry could establish effective financial incentives for encouraging domestic and business consumers to reduce energy consumption. The Treasury could impose a tax on energy consumption and the power supply industry could reform pricing structures to the same end.

Then there are the material checks that could be placed on the domestic and business use of electricity. A state-sponsored Energy Conservation Policy could end the tradition of the inhabitants of our damp and chilly islands letting millions of units of electricity go to waste through lack of proper thermal efficiency in homes and factories. The chief UK legislation currently concerned with improving domestic energy efficiency is the Home Energy Conservation Act that came into force in Scotland at the end of 1996. The Act made local authorities responsible for securing 'substantial' gains in domestic energy efficiency. At the start of the Act's operation, the then Secretary of State for Scotland set an aspirational target of a 30% improvement in Scotland's domestic energy efficiency. The Scottish Executive is now in charge of monitoring Scotland's progress towards this target and the news is not good: an estimated figure of 4.2% improvement in domestic energy efficiency was reported for 2002.

HECA's failure to impact decisively on the thermal efficiency of Scottish homes only confirms the country's need for a realistic, properly resourced and effectively authorised energy policy. HECA's objective was always unlikely to be met because of the insufficient funds made available to the Local Authorities in charge of its execution. This lack of resources limited the number of improvements to energy efficiency that HECA officers were able to carry out. HECA gave Local Authorities no compulsory powers to assess or take action on the domestic energy efficiency status of private dwellings. Local Authorities were authorised to make improvements in their own housing stock but not to gather detailed information about energy efficiency in the private sector. Some funding was made available for energy efficiency improvement grants and publicity schemes in the private sector but it was insufficient and sporadic with schemes like ReWarm being abandoned due to lack of money. Duplication of effort and lack of co-ordination between the many different schemes and agencies involved also hampered the success of HECA.

Indeed, this sorry catalogue of ineffectiveness, covering both supply and demand in the UK and Scottish energy regime, points to the conclusion that anything less than a comprehensive, coordinated and properly funded energy strategy is a waste of valuable time and resources. The environmental crisis threatened by climate change presents society with a vital opportunity to take action; but it must be action appropriately planned and resourced. To this end the creation of an Energy Ministry with adequate political power to prosecute its policies and enough resources to carry them out is indispensable. Already the Confederation of British Industry has complained that European Union regulation of carbon emissions is restricting UK competitiveness. The Confederation would no doubt accuse a worthwhile Energy Ministry of constituting the same sort of drag on economic activity. It would envisage such comprehensive intervention as an inevitable cause of falling exports and rising unemployment: obviously such a policy would never work.

But something like the sort of Energy Ministry we have been imagining does exist in Denmark. Perched on the north west periphery of Europe with a population of just over 5 million, Denmark has been pursuing environment led policies since the

1970s. It was then that the country assumed its position at the forefront of those industrialised countries attempting to face up meaningfully to the challenges posed by that decade's oil crises.

The Danes' keenness to grasp the nettle was at least partly inspired by her own meagre fossil fuel reserves. But, like Scotland, Denmark had a tradition of renewable energy development. In the same decade that the British Aluminium Company built their hydro-powered factory at Foyers, the Danish government funded the design and construction of an experimental windmill. The strategic importance of wind power for the Danes only intensified during the exigencies of two world wars. So necessity and tradition have combined to inspire Denmark's integrated, focused and comprehensive long-term energy policy. Thanks to a thriving wind power sector, which generates 15% of the national power supply, the Danes have met their Kyoto targets already. State support for renewable energy in Denmark meant investment in the manufacture of the specialist equipment used by the generators of all forms of renewable energy.

But support for renewables is only one strand of Danish energy policy. In the 1970s, it was decided that one remedy for fuel shortages would be to encourage reduced consumption of energy. On the assumption that energy efficiency would profit individuals, families, businesses and government, energy saving became compulsory in Denmark. In 1980, the Folketing (the Danish Parliament) gave the country's 275 municipalities the power to design, resource and carry out their own energy saving plans. They set up local energy saving committees to provide information and guidance on energy conservation and generally to oversee the national energy efficiency campaign. The energy efficiency of all buildings in Denmark is now officially audited and must be disclosed to a potential buyer.

The Danish government also continues its support for renewables by investing in research and development. Environmental considerations have inspired Danish waste disposal policies too. Landfill is no longer permitted and waste, which can't be re-used or recycled, is incinerated in Combined Heat and Power plants.

Denmark provides a good example of a country making radical changes for the sake of the environment. The results speak for themselves. Since 1992, Denmark has reduced carbon emissions by about 12%, energy consumption remains stable and GDP has risen, with renewables technology a leading growth sector. The Danish experience would seem to confirm the wisdom of letting environmental priorities rule. HECA looks very half-hearted in comparison to the Danish statutory energy conservation regime.

Moreover, failure to take proper energy conservation seriously is not just environmentally short sighted but condemns poorer households to cold and unhealthy homes that would be illegal in Denmark. Instead of energy policy being just a revenue guarantee for private utilities it should become more of a tool for environmental survival and social justice and one which is pledged to make the most of our own indigenous renewable resources. We vitally need new thinking on energy but the principles of the NOSHEB and the political resolve of its founder, Thomas Johnston, must not be forgotten.

CHAPTER 9

How Green is Hydro-electric Power?

HYDRO-ELECTRIC POWER, whoever is in charge of its development, still attracts fierce opposition from environmentalists, as the Assynt Crofters found out to their cost. Can hydro developers really claim to be working for the environment, or were all hydro's original critics correct in their condemnation of water power as unnatural, unsightly and economically impractical? In other words, is hydro power Green or not? Can it make a contribution to a truly sustainable energy strategy?

To answer this last and most important question about hydro power, I intend to do what politicians and their economist accountants have been doing all through this narrative and move the accounting goalposts. National economic policy considerations were used from the 1950s onward to challenge and eventually undermine state sponsorship of hydro development. As already noted, the accounting methods insisted on by the Treasury during the Mackenzie Inquiry dealt the NOSHEB Development Plan a deadly blow. The Board's building ambitions ended forever in 1983 when its Grudie-Talladale proposals were rejected on economic grounds, as dictated by the Public Sector Borrowing Requirement. However, the realisation is growing across the globe that the priorities and criteria on which traditional economics is based may well be obsolete.

Over the last hundred years, the well-being of any society has been equated with its standard of economic growth, as measured by Gross Domestic Product. However, it is becoming increasingly obvious that the quantity of economic growth alone does not necessarily contribute to our well-being. GDP is calculated in a way that takes no account of the quality of that economic growth. GDP, as worked out at present, makes no distinction between different types of economic activity. All economic activity is counted, regardless of its quality. Symptoms of social disintegration which generate economic activity like accidents, divorce and crime are identified as economic gains. Any attempt to improve our quality of life without endangering our planetary resources requires a new economic model, shaped by different indicators. Groups including Friends of the Earth and the New Economics Foundation have suggested alternative economic indicators like air and water quality, environmental damage and the depletion of environmental assets. The ultimate purpose of this paradigm revision is an attempt to re-set the course of our progress towards the sort of development which will not deprive future generations by squandering environmental assets now.

So how does hydro power fare under this sort of scrutiny? Does it have a significant part to play in a sustainable future? No story of hydro-electric power would be truly complete without an assessment of its future value based not on GDP but, rather, on the sort of environmental audit which deals with sustainability and true ecological value. Our last look at hydro power takes the form of two such environmental audits. The first of these audits focuses on the original NOSHEB project, subsequently managed by Scottish Hydro-Electric plc and now by Scottish and Southern Energy plc. The second of these environmental audits looks at hydro power's global role in sustainable power generation.

An ideal starting point for both audits is a definition of hydro as: 'a fully renewable, benign and proven source of indigenous power with no waste products and no health hazards to its operators or the public'. (Frank Johnson quoted in *The Hydro*, by Peter Payne).

The plant installed across the Highlands during the execution of the NOSHEB Development Plan seems to have stood the test of the last half-century most impressively. All the generating apparatus from tunnels to turbines has been constantly maintained and refurbished ever since it was commissioned, a regime that has raised levels of generating output and efficiency throughout the system.

Sloy, the NOSHEB's first big success story, was upgraded as part of this policy in 1999. Nearly 50 years since it started generating, the entire system at Sloy was overhauled and updated at a cost to Scottish and Southern Energy of £113m. As well as raising operational standards at Sloy, the refurbishments of the turbines, generators, cabling and pipework will extend the operating life of the station for another 30 to 40 years and increase its output by over 20%.

Throughout the system, the dams have been kept in optimum operational condition, although some like Mullardoch have lost their original immaculate appearance to black lichen growth. None of the reservoirs has silted up and the entire 200 miles of the tunnel network has required only routine maintenance. The original NOSHEB water-carrying system has proved less durable above ground, where many exposed concrete pipelines have been corroded by the Highland weather. Some of these aqueducts have had to be re-lined and some replaced.

The NOSHEB and its successors have claimed a good flood control record. And though there may well be more than a few local objections to this boast, the wet Highland winters of the past five decades have not caused flooding of any magnitude.

From the outset, one of the most powerfully-argued environmental objections to hydro power development was the threat that it posed to Highland salmon fisheries. We have seen what extensive measures were taken by hydro developers and their designers to accommodate the concerns of fishing interests. But despite all the fish-counting technology in place on hydro rivers in the north, it is impossible to quantify accurately the effect on fish stocks of hydro power development. So many other factors may have contributed to the disastrous decline in salmon numbers over the last decade, including acid rain and deep sea salmon catches. It seems reasonable enough to suppose that salmon might well not flourish where the natural flow of water has been reduced. However, Highland rivers like the Ewe and the Hope, which have never been used for hydro purposes, have suffered patterns of decline similar to those recorded on hydro rivers over this period. There are evidently a number of contributing factors at play, of which hydro development is only one.

The NOSHEB and its successors have been officially committed to protecting the environment in which their apparatus is installed. Schemes are operated in such a way as to minimise variation in the water levels of storage reservoirs thus preventing the development of extended, sterile foreshores. Power stations have been carefully designed and screened by trees where possible.

Since the late 1950s, technological advances allowed power stations to be built underground; they really are invisible.

And while we are reflecting on the operational achievements of Highland hydro power, we must not forget the engineers and linesmen who erected and maintained the electricity distribution and transmission network in the north, frequently in the teeth of horrendous weather conditions. Tom Johnston's magnificent dream could never have come true without the Herculean efforts of these men.

Pylons have never earned the forgiveness of the environmentalists. Even Sir Edward MacColl was teased by his intellectual friends for the striding armies of pylons necessary for the extension of the National Grid to all parts of the Highlands. More recently, it has become clear that planning regulations should insist, as they do in Canada, America and many European countries, that no-one be allowed to build houses too near to pylons, or put up pylons too near houses. However, if a line of pylons means that it is not necessary to wash my clothes by hand or read by candlelight, I can certainly bear looking at them. Pylons are signs that the landscape can support a fairly comfortable human existence as part of its biodiversity.

The same sort of argument applies to the issue of hydro power's effects on the Highland landscape. Preservation of wilderness scenery may prove to be an impossible luxury on this crowded, power-hungry planet. If power is needed to run society, then what is more sensible than to use sustainable resources to obtain that power? The onus must be on government to promote the use of renewable energy resources as far as possible to reduce the environmental impact of burning fossil fuels.

However, renewable energy must be used in the way that is least disruptive to humans and the environment. As far as hydro is concerned, such sensitive use of the resource must certainly include the development of small-scale, community based schemes. These are ideal for some remote Highland communities. Such communities can benefit greatly from the chance to earn income derived from resources close to hand.

The planning difficulties experienced by the Assynt Crofters suggest that preoccupations with environmental conservation may prove a serious obstacle to community projects. People will have to decide which environment it is that they want to protect: today's or tomorrow's? Landscape is a temporary thing. No place on earth is as it was at the start. The NoSHEB project has affected the Highland landscape, its rivers and lochs. But without the project there might be hardly a soul in that landscape today.

The environmental record of hydro power's operations in the north may be generally good, but when we attempt an environmental audit of hydro power as a global renewable resource this question of environmental impact might be harder to settle. The controversial scale of the Three Gorges hydro power project currently under construction in China brings the question of environmental impact into sharp focus. The dam will reach 1.4 miles across the Yangtse River, stand 604 feet high and create a reservoir 34 miles long. The project has been under consideration since 1919, when it was first proposed by the warlord, Sun Yat Sen. He stressed the value of such a scheme for controlling the Yangtse's tendency to frequent and violent floods. Flood control is still a motivation for the project today.

Critics of the project point to the damage that the massive installation will wreak on the natural and human environment. The development threatens several plant and animal species including the rare Yangtse River Dolphin. There are also fears that industrial and domestic pollutants will form dangerous concentrations if they are not washed out to sea as happens at present. But the most staggering estimate of disruption calculates that the number of people who will face relocation because of inundation could be as many as 1.9 million. These displaced persons will have to get used to much bigger changes than the erection of a few pylons in the next field. Nonetheless, the dam, when it is commissioned in 2009, will be able to provide at least 9% of all China's power needs. It will save our biosphere the huge burden of CO_2 emissions which would be produced if the same energy were generated from fossil fuels. Chinese coal has a high sulphur content which makes its emissions particularly damaging.

Any environmental audit of global hydro power has to deal with some highly controversial pros and cons. But if the ever-increasing human race insists on using more and more power, it may have to tolerate some re-arrangements. The global demand for power is currently increasing by 2% every year. Sustainable solutions to the problems posed by global appetites will have to be a part of any meaningful new world order.

Chronology

1882	Sir William Armstrong builds the first hydro-electric installation in Britain at Cragside House in Northumberland
1890	The private hydro installation at Fort Augustus Abbey is extended to the village to become Scotland's first hydro-powered public electricity supply scheme
1896	British Aluminium Company starts smelting operations at Foyers
1903	Ravensrock Power Station commissioned at Strathpeffer
1909	BAC starts smelting at Kinlochleven
1920	Publication of the Snell Report
1924-27	Clyde Valley Electric Company opens hydro-electric generating scheme with power stations at Stonebyres and Bonnington
1926	The Electricity Act and the beginning of the National Grid
1929	Ross-shire Electric Supply Company opens hydro scheme on Loch Luichart
1930-33	Grampian Electric Supply Company opens hydro scheme with power stations at Rannoch and Tummel Bridge
1934	BAC begins smelting at Fort William
1935-36	Galloway Water Power Company opens hydro scheme with power stations at Tongland (1935), Kendoon (1936), Carsfad (1936) and Earlstoun (1936)
1941	Cooper Committee appointed to investigate the prospects for water power development in the Highlands
1943	Hydro-Electric Development (Scotland) Act and the founding of the North of Scotland Hydro-Electric Board
1944-75	The construction of the NOSHEB Development Plan

1948	Morar	Morar Project
1948	Nostie Bridge	Lochalsh Project
1950	Clunie	Tummel-Garry Project
1950	Grudie Bridge	Fannich Project

1950	Pitlochry	Tummel-Garry Project
1950	Sloy	Sloy Project
1951	Fasnakyle	Affric Project
1951	Kerry Falls	Gairloch Project
1959	Striven	Cowal Project
1952	Storr Lochs	Skye Project
1952	Lussa	Kintyre Project
1952	Gaur	Gaur Project
1954	Luichart	Conon Valley Project
1954	Torr Achilty	Conon Valley Project
1954	Loch Dubh	Ullapool Project
1955	Clachan	Shira Project
1955-6	Errochty	Tummel-Garry Project
1955	Finlarig	Lawers Project
1955	Quoich[Glenmoriston]	Garry Project
1956	Invergarry	Garry Project
1956	Ceannacroc	Moriston Project
1956	Allt-na-Lairige	Lairige Project
1956	Kilmelfort	Kilmelfort Project
1956	Achanalt	Conon Valley Project
1957	Mossford	Conon Valley Project
1957	Ston Mor	Shira Project
1957	St Fillans	Breadalbane Project
1958	Dalchonzie	Breadalbane Project
1958	Lubreoch	Breadalbane Project
1958	Glenmoriston	Moriston Project
1958	Shin	Shin Project
1958	Lochay	Breadalbane Project
1959	Cashlie	Breadalbane Project
1959	Orrin	Conon Valley Project
1967	Cruachan	Awe Project
1975	Foyers	
1981	Alcan refurbishes old BAC smelter at Fort William retaining original hydro power generating components	
1989	Electricity Privatisation	
1990	NOSHEB privatised as Hydro-Electric plc	
1992	Rio Earth Summit	
1993	Hydro-Electric plc becomes Scottish Hydro plc	

1997	Kyoto Conference on Climate Change
1998	Scottish Hydro plc merges with Southern Electric to become Scottish and Southern Energy plc
1999	Scottish Parliament's Cross-Party Group on Renewable Energy established
2000	Kinlochleven smelter closed, being too small to be internationally competitive Scottish and Southern Energy begins construction of hydro installation at Culeig near Ullapool
2001	Passage of the Utilities Act encourages the use of renewable energy to help meet revised UK carbon dioxide emissions targets
2002	Earth Summit on Sustainable Development, Johannesburg
2002	Scottish and Southern Energy plc hydro development at Cuileig completed
2002	UK Energy Review
2004	Scottish and Southern Energy plc hydro development at Kingairloch due to be completed
2004	Scottish Parliament's Enterprise and Culture Committee's Inquiry into Scottish Renewables
2020	(Target) Scotland generates 40% of its electricity from renewable resources
2025*	UK becomes net importer of gas
2060*	Decommissioning and site remediation at Dounreay complete at projected cost of £4 billion
2100*	Exhaustion of known world oil reserves

* Projected Date

Bibliography

Agnew, PW, 1994, *The Efficient Alternative*, Tarragon Press, Glasgow

Ash, Marinell, 1991, *This Noble Harbour*, Invergordon Port Authority/John Donald, Edinburgh

Campbell, Patrick, 2000, *Tunnel Tigers*, PH Campbell, Jersey City.

Collier, Dr Ute, 1994, *Energy and Environment in the European Union*, Avebury Publishing, Aldershot

Department of Trade and Industry, *The Energy Report 2000*

Douglas, Tom, May 1997, *Opportunity Knocks for Scottish Hydro, International Water Power and Dam Construction*, Wilmington Publishing Ltd., Dartford

Eksteins, Modris, 1999, *Walking Since Daybreak*, Houghton Mifflin, New York

Galbraith, Russell, 1995, *Without Quarter*, Mainstream, Edinburgh

Gunn, Neil, 1938, *Off in a Boat*, Faber and Faber, London

Fraser, Norrie (ed.), 1956, *Sir Edward MacColl: a Maker of Modern Scotland*, The Stanley Press, Edinburgh

Friends of the Earth Scotland, 2003, *Warmer Homes, Cooler Planet*

Johnston, Thomas, 1908, *Our Scots Noble Families*, Forward Publishing, Glasgow

Johnston, Thomas, 1920, *History of the Working Class in Scotland*, Forward Publishing, Glasgow

Johnston, Thomas, 1952, *Memories*, Collins, London

Handley, JE, 1970, *The Navvy in Scotland*, Cork University Press, Cork

Hunter, James, 1999, *Last of the Free*, Mainstream, Edinburgh

MacGill, Patrick, 1914, *Children of the Dead End*, Herbert Jenkins, London

Oakley, CA, 1967, *The Second City*, Blackie, Glasgow

Payne, Peter, 1988, *The Hydro*, Aberdeen University Press, Aberdeen

Paton, TAL and Guthrie Brown, J, 1961, *Power from Water*, Leonard Hill, London

Performance and Innovation Unit (Cabinet Office) Report, 2002, *The Energy Review*

Ramage, Janet, 1997, *Energy, A Guidebook*, Oxford University Press, Oxford

Richards, Eric and Clough, Monica, 1989, *Cromartie, a Highland Life*, Aberdeen University Press, Aberdeen

Roy, Arundhati, 1999, *The Cost of Living*, Flamingo, London

Stott, Louis, 1987, *The Waterfalls of Scotland*, Aberdeen University Press

Surrey, John, (ed) 1996, *The British Electricity Experiment*, Earthscan Publications, London

Thomson, Iain, 1981, *Isolation Shepherd*, Bidean Books, Beauly

Ward, Colin, 1997, *Reflected in Water*, Cassell, London

USEFUL WEB-SITES

Friends of the Earth (Scotland)	www.foe-scotland.org.uk
Scottish and Southern Energy	www.scottish.southern.co.uk
Scottish Renewables Forum	www.scottishrenewables.com
British Hydropower Association	www.british-hydro.org
New Economics Foundation	www.neweconomics.org
Earth Summit 2002	www.earthsummit2002.org
Energy Action Scotland	www.eas.org.uk
Scottish Parliament	www.scottish.parliament.uk
Association for the Conservation of Energy	www.ukace.org

Index

Some other books published by **LUATH** PRESS

Notes from the North incorporating a brief history of the Scots and the English

Emma Wood

ISBN 1 84282 048 6 PB £7.99

Notes on being English
Notes on being in Scotland
Learning from a shared past

Sickened by the English jingoism that surfaced in rampant form during the 1982 Falklands War, Emma Wood started to dream of moving from her home in East Anglia to the Highlands of Scotland.

She felt increasingly frustrated and marginalised as Thatcherism got a grip on the southern English psyche. The Scots she met on frequent holidays in the Highlands had no truck with Thatcherism, and she felt at home with grass-roots Scottish anti-authoritarianism. The decision was made. She uprooted and headed for a new life in the north of Scotland.

She was to discover that she had crossed a border in more than the geographical sense. In this book she sets a study of Scots-English conflicts alongside personal experiences of contemporary incomers' lives in the Highlands. Her own approach has been thoughtful and creative. Notes from the North is a pragmatic, positive and forward-looking contribution to cultural and political debate within Scotland.

... her enlightenment is evident on every page of this perceptive, provocative book

MAIL ON SUNDAY

An intelligent and perceptive book... calm, reflective, witty and sensitive. It should certainly be read by all English visitors to Scotland, be they tourists or incomers. And it should certainly be read by all Scots concerned about what kind of nation we live in.

THE HERALD

Scotlands of the Future: sustainability in a small nation

Introduced and edited by Eurig Scandrett

ISBN 1 84282 035 4 PB £7.99

What sorts of futures are possible for Scotland?

How can citizens of a small nation at the periphery of the global economy make a difference?

Can Scotland's economy be sustainable?

How do we build a good quality of life without damaging others'?

Could there be an economy that is good for people and the environment?

And if so, how do we get there without damaging people's livelihoods?

What can the Scottish Parliament do?

What difference can we make in our organisations, our trade unions, and our businesses?

Scotlands of the Future looks at where we've got to, where we can go next, and where we might want to get to – essential reading for those who think about and want to take action for a sustainable Scotland, and anyone else who cares about the future.

The anti-slavery campaigners succeeded. Politicians and civil society must rise to this new challenge, which is just today's version of the same injustice. We must show imagination, courage and leadership and champion a sustainable economy – for Scotland and the world.

OSBERT LANCASTER, EXECUTIVE DIRECTOR, CENTRE FOR HUMAN ECOLOGY

Shale Voices
Alistair Findlay
foreword by Tam Dalyell MP
ISBN 0 946487 78 2 HB £17.99
ISBN 0 946487 63 4 PB £10.99

He was at Addiewell oil works. Anyone goes in there is there for keeps.
JOE, Electrician

There's shale from here to Ayr, you see.
DICK, a Drawer

The way I describe it is, you're a coal miner and I'm a shale miner. You're a tramp and I'm a toff.
HARRY, a Drawer

There were sixteen or eighteen Simpsons...
...She was having one every dividend we would say.
SISTERS, from Broxburn

Shale Voices offers a fascinating insight into shale mining, an industry that employed generations of Scots, had an impact on the social, political and cultural history of Scotland and gave birth to today's large oil companies. Author Alistair Findlay was born in the shale mining village of Winchburgh and is the fourth son of a shale miner, Bob Findlay, who became editor of the *West Lothian Courier*. *Shale Voices* combines oral history, local journalism and family history. The generations of communities involved in shale mining provide, in their own words, a unique documentation of the industry and its cultural and political impact.

Photographs, drawings, poetry and short stories make this a thought provoking and entertaining account. It is as much a joy to dip into and feast the eyes on as to read from cover to cover.

Alistair Findlay has added a basic source material to the study of Scottish history that is invaluable and will be of great benefit to future generations. Scotland owes him a debt of gratitude for undertaking this work. TAM DALYELL MP

Scotland - Land and Power the agenda for land reform
Andy Wightman
in association with
Democratic Left Scotland
foreword by Lesley Riddoch
ISBN 0 946487 70 7 PB £5.00

What is land reform?
Why is it needed?
Will the Scottish Parliament really make a difference?

Scotland – Land and Power argues passionately that nothing less than a radical, comprehensive programme of land reform can make the difference that is needed. Now is no time for palliative solutions which treat the symptoms and not the causes.

Scotland – Land and Power is a controversial and provocative book that clarifies the complexities of landownership in Scotland. Andy Wightman explodes the myth that land issues are relevant only to the far flung fringes of rural Scotland, and questions mainstream political commitment to land reform. He presents his own far-reaching programme for change and a pragmatic, inspiring vision of how Scotland can move from outmoded, unjust power structures towards a more equitable landowning democracy.

Writers like Andy Wightman are determined to make sure that the hurt of the last century is not compounded by a rushed solution in the next. This accessible, comprehensive but passionately argued book is quite simply essential reading and perfectly timed – here's hoping Scotland's legislators agree.
LESLEY RIDDOCH

Old Scotland New Scotland

Jeff Fallow

ISBN 0 946487 40 5 PB £6.99

Together we can build a new Scotland based on Labour's values. DONALD DEWAR, Party Political Broadcast

Despite the efforts of decent Mr Dewar, the voters may yet conclude they are looking at the same old hacks in brand new suits.

IAN BELL, THE INDEPENDENT

At times like this you suddenly realise how dangerous the neglect of Scottish history in our schools and universities may turn out to be.

MICHAEL FRY, THE HERALD

...one of the things I hope will go is our chip on the shoulder about the English... The SNP has a huge responsibility to articulate Scottish independence in a way that is pro-Scottish and not anti-English. ALEX SALMOND, THE SCOTSMAN

Scottish politics have never been more exciting. In *Old Scotland New Scotland* Jeff Fallow takes us on a graphic voyage through Scotland's turbulent history, from earliest times through to the present day and beyond. This fast-track guide is the quick way to learn what your history teacher didn't tell you, essential reading for all who seek an understanding of Scotland and its history.

Eschewing the romanticisation of his country's past, Fallow offers a new perspective on an old nation.

Too many people associate Scottish history with tartan trivia or outworn romantic myth. This book aims to blast that stubborn idea. JEFF FALLOW

Eurovision or American Dream? Britain, the Euro and the Future of Europe

David Purdy

ISBN 1 84282 036 2 PB £3.99

Should Britain join the euro?

Where is the European Union going?

Must America rule the world?

Eurovision or American Dream? assesses New Labour's prevarications over the euro and the EU's deliberations about its future against the background of transatlantic discord. Highlighting the contrasts between European social capitalism and American free market individualism, David Purdy shows how Old Europe's welfare states can be renewed in the age of the global market. This, he argues, is essential if European governments are to reconnect with their citizens and revive enthusiasm for the European project. It would also enable the EU to challenge US hegemony, not by transforming itself into a rival superpower, but by championing an alternative model of social development and changing the rules of the global game.

In this timely and important book David Purdy explains why joining the euro is not just a question of economics, but a question about the future political direction of Britain and its place in Europe. PROFESSOR ANDREW GAMBLE, DIRECTOR: POLITICAL ECONOMY RESEARCH CENTRE, DEPARTMENT OF POLITICS, UNIVERSITY OF SHEFFIELD

But n Ben A-Go-Go

Matthew Fitt

ISBN 0 946487 82 0 HB £10.99
ISBN 1 84282 041 1 PB £6.99

The year is 2090. Global flooding has left most of Scotland under water. The descendants of those who survived God's Flood live in a community of floating island parishes, known collectively as Port. Port's citizens live in mortal fear of Senga, a supervirus whose victims are kept in a giant hospital warehouse in sealed capsules called Kists. Paolo Broon is a low-ranking cyberjanny. His life-partner, Nadia, lies forgotten and alone in Omega Kist 624 in the Rigo Imbeki Medical Center. When he receives an unexpected message from his radge criminal father to meet him at But n Ben A-Go-Go, Paolo's life is changed forever. He must traverse VINE, Port and the Drylands and deal with rebel American tourists and crabbit Dundonian microchips to discover the truth about his family's past in order to free Nadia from the sair grip of the merciless Senga. Set in a distinctly unbonnie future-Scotland, the novel's dangerous atmosphere and psychologically-malkied characters weave a tale that both chills and intrigues. In But n Ben A-Go-Go Matthew Fitt takes the allegedly dead language of Scots and energises it with a narrative that crackles and fizzes with life.

an entertaining and ground-breaking book
EDWIN MORGAN

... if you can't get hold of a copy, mug somebody
MARK STEPHEN, SCOTTISH CONNECTION, BBC RADIO SCOTLAND

...the last man who tried anything like this was Hugh MacDiarmid MICHAEL FRY, TODAY PROGRAMME, BBC RADIO 4

Bursting with sly humour, staggeringly imaginative, often poignant and at times exploding with Uzi-blazing action, this book is a cracker... With Matthew Fitt's book I began to think and sometimes dream in Scots. GREGOR STEELE, TIMES EDUCATIONAL SUPPLEMENT

The Road Dance

John MacKay

ISBN 1 84282 040 0 PB £6.99

Why would a young woman, dreaming of a new life in America, sacrifice all and commit an act so terrible that she severs all hope of happiness again?

Life in the Scottish Hebrides can be harsh – 'The Edge of the World' some call it. For Kirsty MacLeod, the love of Murdo and their dreams of America promise an escape from the scrape of the land, the suppression of the church and the inevitability of the path their lives would take.

But the Great War looms and Murdo is conscripted. The villagers hold a grand Road Dance to send their young men off to battle. As the dancers swirl and sup, the wheels of tragedy are set in motion.

[MacKay] has captured time, place and atmosphere superbly... a very good debut
MEG HENDERSON

Powerful, shocking, heartbreaking...
DAILY MAIL

With a gripping plot that subtly twists and turns, vivid characterisation and a real sense of time and tradition, this is an absorbing, powerful first novel. The impression it made on me will remain for some time.
THE SCOTS MAGAZINE

SOCIAL HISTORY

Pumpherston: the story of a shale oil village
Sybil Cavanagh
ISBN 1 84282 011 7 HB £17.99
ISBN 1 84282 015 X PB £10.99

Crofting Years
Francis Thompson
ISBN 0 946487 06 5 PB £6.95

A Word for Scotland
Jack Campbell
ISBN 0 946487 48 0 PB £12.99

HISTORY

Scots in Canada
Jenni Calder
ISBN 1 84282 038 9 PB £7.99

Plaids & Bandanas: Highland Drover to Wild West Cowboy
Rob Gibson
ISBN 0 946487 88 X PB £7.99

A Passion for Scotland
David R Ross
ISBN 1 84282 019 2 PB £5.99

Civil Warrior: extraordinary life & poems of Montrose
Robin Bell
ISBN 184282 013 3 HB £10.99

Reportage Scotland: History in the Making
Louise Yeoman
ISBN 0 946487 43 X PB £9.99

NATURAL WORLD

Red Sky at Night
John Barrington
ISBN 0 946487 60 X PB £8.99

The Highland Geology Trail
John L Roberts
ISBN 0 946487 36 7 PB £4.99

Wild Lives: Otters – On the Swirl of the Tide
Bridget MacCaskill
ISBN 0 946487 67 7 PB £9.99

Wild Lives: Foxes – The Blood is Wild
Bridget MacCaskill
ISBN 0 946487 71 5 PB £9.99

Listen to the Trees
Don MacCaskill
ISBN 0 946487 65 0 PB £9.99

Scotland – Land & People: An Inhabited Solitude
James McCarthy
ISBN 0 946487 57 X PB £7.99

BIOGRAPHY

Tobermory Teuchter
Peter Mcnab
ISBN 0 946487 41 3 PB £7.99

Bare Feet and Tackety Boots
Archie Cameron
ISBN 0 946487 17 0 PB £7.95

The Last Lighthouse
Sharma Krauskopf
ISBN 0 946487 96 0 PB £7.99

Come Dungeons Dark
John Taylor Caldwell
ISBN 0 946487 19 7 PB £6.95

LUATH GUIDES TO SCOTLAND

Mull & Iona: Highways and Byways
Peter Mcnab
ISBN 0 946487 58 8 PB £4.95

The North West Highlands: Roads to the Isles
Tom Atkinson
ISBN 0 946487 54 5 PB £4.95

The Northern Highlands: The Empty Lands
Tom Atkinson
ISBN 0 946487 55 3 PB £4.95

The West Highlands: The Lonely Lands
Tom Atkinson
ISBN 0 946487 56 1 PB £4.95

South West Scotland
Tom Atkinson
ISBN 0 946487 04 9 PB £4.95

TRAVEL AND LEISURE

Pilgrims in the Rough: St Andrews Beyond the 19th Hole
Michael Tobert
ISBN 0 946487 74 X PB

Die Kleine Schottlandfibel (Scotland Guide in German)
Hans-Walter Arends
ISBN 0 946487 89 8 PB £8.99

Edinburgh's Historic Mile
Duncan Priddle
ISBN 0 946487 97 9 PB £2.99

POLITICS & CURRENT ISSUES

Scotlands of the Mind
Angus Calder
ISBN 1 84282 008 7 PB £9.99

Trident on Trial: The Case for People's Disarmament
Angie Zelter
ISBN 1 84282 004 4 PB £9.99

Uncomfortably Numb: A Prison Requiem
Maureen Maguire
ISBN 1 84282 001 X PB £8.99

Some Assembly Required: Scottish Parliament
David Shepherd
ISBN 0 946487 84 7 PB £7.99

ISLANDS

The Islands that Roofed the World: Easdale, Belnahua, Luing & Seil
Mary Withall
ISBN 0 946487 76 6 PB £4.99

Rum: Nature's Island
Magnus Magnusson
ISBN 0 946487 32 4 PB £7.95

THE QUEST FOR

The Quest for Robert Louis Stevenson
John Cairney
ISBN 0 946487 87 1 HB £16.99

The Quest for the Nine Maidens
Stuart McHardy
ISBN 0 946487 66 9 HB £16.99

The Quest for the Original Horse Whisperers
Russell Lyon
ISBN 1 84282 020 6 HB £16.99

The Quest for the Celtic Key
Karen Ralls-MacLeod and Ian Robertson
ISBN 1 84282 031 1 PB £8.99

The Quest for Arthur
Stuart McHardy
ISBN 1 84282 012 5 HB £16.99

ON THE TRAIL OF

On the Trail of William Wallace
David R Ross
ISBN 0 946487 47 2 PB £7.99

On the Trail of Bonnie Prince Charlie
David R Ross
ISBN 0 946487 68 5 PB £7.99

On the Trail of Rob Roy MacGregor
John Barrington
ISBN 0 946487 59 6 PB £7.99

On the Trail of Queen Victoria in the Highlands
Ian R Mitchell
ISBN 0 946487 79 0 PB £7.99

On the Trail of John Muir
Cherry Good
ISBN 0 946487 62 6 PB £7.99

On the Trail of Robert the Bruce
David R Ross
ISBN 0 946487 52 9 PB £7.99

On the Trail of Mary Queen of Scots
J Keith Cheetham
ISBN 0 946487 50 2 PB £7.99

FICTION

Outlandish Affairs: An Anthology of Amorous Encounters
Edited and introduced by Evan Rosenthal and Amanda Robinson
ISBN 1 84282 055 9 PB £9.99

Driftnet
Lin Anderson
ISBN 1 84282 034 6 PB £9.99

The Bannockburn Years
William Scott
ISBN 0 946487 34 0 PB £7.95

The Fundamentals of New Caledonia
David Nicol
ISBN 0 946487 93 6 HB £16.99

Milk Treading
Nick Smith
ISBN 1 84282 037 0 PB £6.99

The Strange Case of RL Stevenson
Richard Woodhead
ISBN 0 946487 86 3 HB £16.99

Grave Robbers
Robin Mitchell
ISBN 0 946487 72 3 PB £7.99

The Great Melnikov
Hugh MacLachlan
ISBN 0 946487 42 1 PB £7.95

POETRY

Tartan and Turban
Bashabi Fraser
ISBN 1 84282 044 3 PB £8.99

Drink the Green Fairy
Brian Whittingham
ISBN 1 84282 020 6 PB £8.99

The Ruba'iyat of Omar Khayyam, in Scots
Rab Wilson
ISBN 1 84282 046 X PB £8.99

Talking with Tongues
Brian Finch
ISBN 1 84282 006 0 PB £8.99

Kate o Shanter's Tale and other poems (book)
Matthew Fitt
ISBN 1 84282 028 1 PB £6.99

Kate o Shanter's Tale and other poems (audio CD)
Matthew Fitt
ISBN 1 84282 043 5 £9.99

Immortal Memories: A Compilation of Toasts to the Memory of Burns as delivered at Burns Suppers, 1801-2001
John Cairney
ISBN 1 84282 009 5 HB £20.00

Madame Fifi's Farewell and other poems
Gerry Cambridge
ISBN 1 84282 005 2 PB £8.99

Scots Poems to be Read Aloud
Introduction by Stuart McHardy
ISBN 0 946487 81 2 PB £5.00

Poems to be Read Aloud
Introduction by Tom Atkinson
ISBN 0 946487 006 PB £5.00

Bad Ass Raindrop
Kokumo Rocks
ISBN 1 84292 018 4 PB £6.99

Sex, Death & Football
Alistair Findlay
ISBN 1 84282 022 2 PB £6.99

The Luath Burns Companion
John Cairney
ISBN 1 84282 000 1 PB £10.00

Men and Beasts: Wild Men and Tame Animals
Valerie Gillies and Rebecca Marr
ISBN 0 946487 928 PB £15.00

Blind Harry's Wallace
Hamilton of Gilbertfield (introduced and edited by Elspeth King)
ISBN 0 946 487 43 X HB £15.00
ISBN 0 946487 33 2 PB £8.99

The Whisky Muse: Scotch Whisky in Poem and Song
Robin Laing
ISBN 1 84282 041 9 PB £7.99

WALK WITH LUATH

Mountain Outlaw
Ian R Mitchell
ISBN 1 84282 027 3 PB £6.50

Skye 360: Walking the Coastline
Andrew Dempster
ISBN 0 946487 85 5 PB £8.99

Mountain Days and Bothy Nights
Dave Brown/Ian R Mitchell
ISBN 0 946487 15 4 PB £7.50

Walks in the Cairngorms
Ernest Cross
ISBN 0 946487 09 X PB £4.95

Short Walks in the Cairngorms
Ernest Cross
ISBN 0 946487 23 5 PB £4.95

MUSIC & DANCE, AND WEDDINGS

The Scottish Wedding Book
G Wallace Lockhart
ISBN 1 84282 010 9 PB £12.99

Fiddles and Folk
G Wallace Lockhart
ISBN 0 946487 38 3 PB £7.95

Highland Balls and Village Halls
G Wallace Lockhart
ISBN 0 946487 12 X PB £6.95

FOLKLORE

The Supernatural Highlands
Francis Thompson
ISBN 0 946487 31 6 PB £8.99

Luath Storyteller: Highland Myths & Legends
George W MacPherson
ISBN 1 84282 003 6 PB £5.00

Tales of the North Coast
Alan Temperley
ISBN 0 946487 18 9 PB £8.99

SPORT

Over the Top with the Tartan Army
Andy McArthur
ISBN 0 946487 45 6 PB £7.99

Ski & Snowboard Scotland
Hilary Parke
ISBN 0 946487 35 9 PB £6.99

CARTOONS

Broomie Law
Cinders McLeod
ISBN 0 946487 99 5 PB £4.00

GENEALOGY

Scottish Roots: Step-by-Step Guide for Ancestor Hunters
Alwyn James
ISBN 1 84282 007 9 PB £ 9.99

LANGUAGE

Luath Scots Language Learner (Book)
L Colin Wilson
ISBN 0 946487 91 X PB £9.99

Luath Scots Language Learner (Double Audio CD Set)
L Colin Wilson
ISBN 1 84282 026 5 PB £16.99

HEALTH & SELF HELP

Napiers History of Herbal Healing, Ancient and Modern
Tom Atkinson
ISBN 1 84282 025 7 HB £16.99

Luath Press Limited

committed to publishing well written books worth reading

LUATH PRESS takes its name from Robert Burns, whose little collie Luath (*Gael.*, swift or nimble) tripped up Jean Armour at a wedding and gave him the chance to speak to the woman who was to be his wife and the abiding love of his life. Burns called one of *The Twa Dogs* Luath after Cuchullin's hunting dog in *Ossian's Fingal*. Luath Press was established in 1981 in the heart of Burns country, and is now based a few steps up the road from Burns' first lodgings on Edinburgh's Royal Mile.

Luath offers you distinctive writing with a hint of unexpected pleasures.

Most bookshops in the UK, the US, Canada, Australia, New Zealand and parts of Europe either carry our books in stock or can order them for you. To order direct from us, please send a £sterling cheque, postal order, international money order or your credit card details (number, address of cardholder and expiry date) to us at the address below. Please add post and packing as follows: UK – £1.00 per delivery address; overseas surface mail – £2.50 per delivery address; overseas airmail – £3.50 for the first book to each delivery address, plus £1.00 for each additional book by airmail to the same address. If your order is a gift, we will happily enclose your card or message at no extra charge.

ILLUSTRATION: IAN KELLAS

Luath Press Limited
543/2 Castlehill
The Royal Mile
Edinburgh EH1 2ND
Scotland

Telephone: 0131 225 4326 (24 hours)
Fax: 0131 225 4324
email: gavin.macdougall@luath.co.uk
Website: www.luath.co.uk